Przewalski's Horses in Eurasia

Przewalski's Horses in Eurasia

Pluralism in International Reintroduction Biology

Michael L. Zukosky

LEXINGTON BOOKS
Lanham • Boulder • New York • London

Published by Lexington Books
An imprint of The Rowman & Littlefield Publishing Group, Inc.
4501 Forbes Boulevard, Suite 200, Lanham, Maryland 20706
www.rowman.com

Unit A, Whitacre Mews, 26-34 Stannary Street, London SE11 4AB

British Library Cataloguing in Publication Information Available

Library of Congress Cataloging-in-Publication Data

Names: Zukosky, Michael L., 1973- , author.
Title: Przewalski's horses in Eurasia : pluralism in international reintroduction biology / Michael L. Zukosky.
Description: Lanham : Lexington Books, 2016. | Includes bibliographical references and index.
Identifiers: LCCN 2016032912 (print) | LCCN 2016039656 (ebook) | ISBN 9781498521352 (cloth : alk. paper) | ISBN 9781498521369 (electronic)
Subjects: LCSH: Przewalski's horse—Behavior—Eurasia. | Przewalski's horse—Reintroduction—Mongolia—International cooperation. | Pluralism.
Classification: Classification: LCC QL737.U62 Z85 2016 (print) | LCC QL737.U62 (ebook) | DDC 599.665/5095—dc23
LC record available at https://lccn.loc.gov/2016032912

∞™ The paper used in this publication meets the minimum requirements of American National Standard for Information Sciences Permanence of Paper for Printed Library Materials, ANSI/NISO Z39.48-1992.

Printed in the United States of America

Contents

Preface

This book is about international reintroduction biology. I am a U.S. anthropologist who has researched the relationship between environmental policy and Inner Asian pastoral culture. I first learned about Przewalski's horses, an endangered species currently being reintroduced in northwest China, while working in the region. In this book, I turn my gaze from Chinese grassland science, its role in policy-making, and its impact on pastoral culture to the complexity of international reintroduction biology. For those familiar with my earlier research, I am using a different theoretical framework and approach because the "horses" and "international reintroduction biology" are different phenomena than "grasslands" and "Chinese grasslands science." For me, Przewalski's horse reintroduction required a different way of thinking and making sense of the world.

This book is also different than my earlier research in terms of its methods. The horses are now being reintroduced, among a range of other sites, to the winter pastures of pastoralists in northwestern China. My research has involved consulting for reintroduction biologists themselves, my own independently funded research, and the complex practice of interpretive analysis and explanation. As I engaged the international network of people involved in horse reintroduction, I spent ten years conducting research in China and elsewhere, doing analysis of my data, and reflecting deeply on what I have been learning. The book is the result of such a praxis, moving between participating in reintroduction as a social scientist and the detached and academic reflection of theory construction. Nevertheless, it began with a very applied question related to the impacts of reintroduction activities on local communities of pastoralists, even when the text itself goes quite beyond such a question.

Many anthropologists struggle with situating themselves either with problem-solving practical matters of public concern or issues of pure theory, publication, and teaching. This work is based on the conscious decision to link and move between these domains. It involves both the identification, assessment, and problem-solving that is required by a broader human and non-human community as well as the theoretical explanation and educational work done within academic anthropology. It began as letters and a report to biologists involved in reintroduction, and through my other work as a teaching and research anthropologist, slowly developed into its present form. In trying to understand the experiences of horses, the biologists who study and work with them, and others involved in reintroduction, I explored new theoretical literatures. In an early rough draft, it was assigned to my undergraduate students and was the object of our classroom discussions about how anthropologists describe and explain science as well as the theoretical frameworks that have resulted. The need for social assessments of the activities of scientists has become widespread, and thus, such an academic endeavor is inherently applied, bringing an issue of public concern to young people and future generations through educational institutions like the university. Furthermore, I presented my arguments to my university colleagues, where Science, Technology, Engineering, and Mathematics (STEM) now dominate, and at professional scientific conferences, where new perspectives on science are often, but not always, encouraged and welcomed. These arguments were summarized in a formal letter to the U.S. Supreme Court concerning the significance of racial and ethnic diversity in natural science classrooms and institutions as well as a defense of positive discrimination in university admissions.

Therefore, I hope this work, in a general way, stimulates discussions about the false and problematic dichotomy of applied versus theoretical research. In some ways, it is a product of a traditional applied anthropology where the anthropologist becomes an expert for a community and applies their methods and knowledge to help that community solve a problem. In other ways, it is very traditional academic anthropology that is concerned with explanation, theory construction, and broader issues within education. While it explores basic questions about humans and other species, the text also explores the various implications of these questions for both reintroduction biology, as well as implicitly within a range of STEM programs and policy. The book is ultimately about how we might reframe future approaches to international scientific practice and ultimately do such science better.

Acknowledgments

This book is based upon research supported by the National Science Foundation under Grant No. 0750634. Any opinions, findings, conclusions, and/or recommendations expressed are those of the author and do not necessarily reflect the views of the National Science Foundation. In addition, the American Center for Mongolian Studies and Eastern Washington University also provided financial support. I am grateful for all of the support of these institutions.

Of course, support for this project came from many directions and in many forms. I would like to thank the staff of the Cologne Zoo, the Smithsonian Conservation Biology Institute, Beijing Forestry University, National University of Mongolia, the Wild Horse Breeding Center, and Hustai Nuruu National Park. Other supporting institutions include Altyn-Emez National Park (Kazakhstan), the Beijing Zoo, the National Museum of Natural History (France), and Utrecht University (Netherlands). These institutions were essential in the research for this book, and I deeply appreciate the cooperation that each offered to me. While I am indebted to these institutions for my research, all mistakes and errors are my own.

I would also like to thank the following individuals: Chalkhar Akenbek, B. Baigalmaa, Mikel Bennett, Cao Qing, Cate Christen, Chen Yun, Chen Xiangjun, Kim Crowder, Machteld van Dierendonck, Chimedtseren Dorjgotov, Ashely Gamboa, Chea Hoban, Hu Defu, Petra Kaczensky, Sarah King, Lydia Kolter, Peter Leimgruber, Camille McNeely, Bandi Namkhai, Norbert Pantel, Franziska Roth, Vickie Shields, Melissa Songer, Chris Walzer, Stacy Warren, Yang Liangliang, Yu Xiaojie, Mariam Zaina, and Waltraut Zimmerman. The work of each and every one of you made this book possible and stimulated its growth. I have deeply appreciated your assistance and support.

Finally, I would like to thank my parents as well as my wife and son, Kara and Ziya. You are the greatest joys of my life.

Introduction

Pluralism and the Reintroduction of Przewalski's Horses

Przewalski's horses (*Equus ferus przewalskii*) are a genetically distinct and distant relative of today's domesticated horses. Although they can hybridize with domestic horses to produce fertile offspring, the geneticist Oliver Ryder (1994) has argued that Przewalski's horses are more biologically distinct from their domestic relatives than are any two breeds of domestic horse. They are distinctive in their physical appearance too; shorter and stockier than domestic horses, they have erect manes with no forelock as well as short hairs on the upper part of their tails. In contrast, domestic horses have long, falling manes and long hairs all over their tails. While domestic horses have a great variety of coloring, Przewalski's horses also have a more consistent dun color with a dark dorsal stripe down their back. Using this genetic and morphological evidence, biologists have argued that Przewalski's horses are thus an important, distinct species of horse.

While Przewalski's horses' historic and geographic distribution in the wild previously ranged from Europe to Asia, the horse went extinct in the wild and was last seen in China in the 1950's. In a positive turn of events, the International Union for the Conservation of Nature's Red List of Threatened Species changed the classification of Przewalski's horse from "extinct in the wild" to "critically endangered" in 2008. This was the result of the efforts of an international network of people involved in captive breeding and reintroduction back to the wild. At present, there are around 2000 individual animals remaining both in North American and European zoos as well as in contemporary reintroduction sites in Asia. The work of biologists researching the horses is important because it both helps to maintain the life of this unique species as well as contribute to the maintenance of the general biodiversity of our planet.

Przewalski's horses are one of the many species being reintroduced throughout the world. Species reintroduction and reintroduction biology are contemporary ways that people engage with ecological problems such as intensification of human land use and the resulting fragmentation of wildlife habitat as human land use and related pressures on natural ecosystems are now beyond so-called safe limits (Newbold et al. 2016). The endangerment of many populations of small and large animals is a significant on-going problem, and a number of species are in the process of going extinct. As one example of the human response to this problem, scientists breed and introduce back to their native habitats endangered animal species such as the Przewalski horse. This complex process is commonly known as species reintroduction, and its science is known as reintroduction biology.

This book describes and explains the ways in which pluralism operates within the network of Przewalski horse reintroduction. Pluralism, as I use the concept, is the notion that any one single approach or method of interpretation in understanding complex phenomena like Przewalski horse reintroduction is liable to be insufficient. The pluralism that I explore involves the complex nature of the horses themselves and of those human societies responding to species endangerment. For example, biologists breed Przewalski's horses and then reintroduce individuals and groups to their native habitats across the Eurasian steppe, an ecological area of grassland plains that stretches across many geographical regions, from eastern Europe to east Asia. Within reintroduction, there are different horses reintroduced and observed by different people at various ecological sites, processes which generate and reflect a multitude of complementary and contradictory truths. Various human communities control these reintroduction sites, and each of them is both involved in reintroduction and affected by it in ways that defy any one particular interpretation or conclusion. Understanding both Przewalski horses and Przewalski horse reintroduction today entails understanding the concept of pluralism.

To understand this pluralism, I have interviewed scientists, analyzed their writings, and participated in the process of reintroduction. I have immersed myself in the working lives of scientists and in the lives of other personnel involved with reintroduction at a number of sites throughout the world. I accompanied specialists as they conducted research at breeding centers and nature reserves, read their published articles, and listened to them read unpublished work at scientific conferences. I observed rangers at reintroduction sites as they monitored released horses, spoke with local people who provided assistance to scientists, and interviewed resident citizens and political officials about their perceptions of reintroduction biology. I have also spent time watching the horses themselves. In the pages ahead, I introduce many interesting people (and horses) and tell their diverse stories.

As a scientific discipline, reintroduction biology tends to lay emphasis on technical information, on its own concepts and theories, and on its latest developments and innovations. For example, a German biologist engaged in the captive breeding of Przewalski's horses might study the distinctive genetics of the species and how genetic principles can be used to help establish self-sustaining populations in captivity. She or he might discuss the cutting-edge scientific capabilities of a particular university's research or breeding center. She or he might tell us when and why the species went extinct, or recount attempts at reintroduction to the wild, and discuss the existing condition of released populations. All this information is important and valuable, but it is different from an anthropology of reintroduction biology.

As an anthropologist, I emphasize the philosophical dilemmas of pluralism for scientists' development of an adequate knowledge of nature and a plan of action to resolve species loss. The reintroduction biology that I discuss here conforms neither to an idealized image of science as a single method, a universally valid and singular truth, nor to a disinterested and objective application of such knowledge to nature. In the case of Przewalski horse reintroduction biology, I see scientists struggling to develop knowledge of a complex species that often exceeds their ability to know and control. In their endeavors, they utilize an array of methods and ways of knowing derived from the expertise of nonscientists. As much as they are logical or rational, the practices of reintroduction biology also entail various culturally-specific and contingent logics, values, and even, emotions. Rather than setting out to simply solve problems related to species extinction and human land use, reintroduction biology actually complicates these matters, as human communities both cooperate with scientists as well as challenge them on political, ethical, or other grounds. Beyond simply representing biologists, biological knowledge, or the problems to which it is applied, this book considers these dilemmas and the potential consequences of pluralism for the future of Przewalski horse reintroduction and the scientific practices related to it.

THE ANTHROPOLOGY OF SCIENCE

In the past, anthropologists had an idealized image of both their own scientific discipline as well as others like biology, and their goal was the formulation of a universally valid and singular truth about being human. By the use of the scientific method, past anthropologists attempted to observe, record, and measure human behavior and to establish objective regularities and general laws relating to it. With a particular hypothesis about humans, anthropologists embarked on observing the lives of those they studied, and aimed to validate or refute their own assumptions with the data they collected. This is science as scientists think it should be, an ideal, rather than science as it is.

Anthropologists began to reflect critically on their practice and realized that they produced truths that were also shared and learned, dynamic and changing—and critically, such truths were part of the anthropologists' own culture. A critique of scientific objectivity in anthropology developed out of a reflection on the cultural assumptions in using such scientific methods. By the 1980s, anthropologists began reading the historian of science, Thomas Kuhn (1996 [1962]), who argued that scientists develop theories and explanations of the world, not just by creating hypotheses and testing with different methods, but through what he called "paradigms." Paradigms are shared and learned sets of ideas and practices that allow for the formation of scientific hypotheses and their testing with methods. For example, early twentieth-century paradigms provided anthropologists with hypotheses about human forms of kinship understood as relationships between blood relatives and in-laws. On the basis of these assumptions, a scientist could make particular hypotheses and test them through observations about the world. Yet, the hypotheses themselves came from the paradigm, not from the world itself, and could therefore be understood to be different from that world. They were not simply objective, but instead reflected a particular perspective which was biased towards the norms of the anthropologists' own culture.

The paradigmatic assumptions which are embedded within the questions anthropologists ask of human behavior and the answers such questions allow one to discover both draw on quite particular cultural contexts. A relevant example of this appeared when the emergence of new reproductive technologies in Europe challenged earlier assumptions about sexual reproduction, parenting, and the legal frameworks of marriage. In response to this new cultural phenomenon, the anthropologist Marilyn Strathern (1989; 1992) noted that previous anthropological paradigms were simply reflections of Western cultural concepts of relatedness rather than unproblematic and universal explanations of culturally distinctive kinship forms. With the work of Strathern, anthropologists had clearly come to some important new understandings of the limits of what could be known about humanity through anthropological practice. Firstly, they generally came to realize the power and limits of scientific knowledge. Secondly, they saw the role played by cultural assumptions in the production of scientific knowledge. Thirdly, they apprehended the immense complexity of the physical, material, and human worlds.

Following these realizations, anthropologists began to study how other scientific disciplines embody culture too. These anthropologists used their methods to describe how the assumptions and practices of a science relate to a broader culture (Franklin 1995). For example, Stefan Helmreich (2009) studied marine microbiologists and explored their means of knowing microscopic lifeforms in the ocean. In capturing the cultural context of scientific fact production, Helmreich argues that microbiologists create hypotheses and make observations relative to particular social networks of scientists and nonscientists that include, amongst others, bioprospecting corporations that seek to discover genes which have commercial value in the market economy. In Helmreich's account, producing facts within marine microbiology is related to the production, distribution, and consumption of goods and services of a group of people. Like anthropology, microbiology is not simply about observing, recording, and measuring the world so as to establish regularities and general laws: both disciplines reflect the particular ideas, social relations, and economic practices that we call "culture."

While marine microbiology might be understood as a "culture," scientific practice is diverse and escapes any one simple understanding. While scientific practice might relate to social relations and economic practices generally, the specific relationships and practices involved in any one science are contextually particular. In fact, recent anthropologists are highlighting how each individual science can bring a multiplicity of different phenomena together in its unique practice. A recent collaboration of anthropologists (Choy et al. 2009; Hathaway 2009; 2014; Tsing 2015) looks at the diverse mycologists, forest managers, mushroom pickers, restaurant chefs, and others involved in the global world of Matsutake mushrooms. Together they show how the science of mycology is located at the intersection of many different materialities and knowledges and can embrace social relations amongst co-workers and friends, and unexpected aspects of human life that include aromas, senses of beauty, and a desire for strength. An anthropological scientist might very well assume that the particular aspects of the material world and its intersection with multicultural networks of people will be different, even unpredictable, in other scientific disciplines and collaborations. Science might be cultural, but it is multicultural too.

The idea of cultural pluralism in science was my general theoretical approach when I began my work with Przewalski's horses and reintroduction biologists. By studying German and Chinese scientists managing and researching small horse populations in captivity in Europe and China, I saw how these different groups of scientists need to work together with local communities and governments to restablish wild populations at individual sites in Eurasia, Mongolia, and Hungary. I assumed that international reintroduction biology would be deeply multicultural and would reflect a variety of ideas about the material world, different relationships between different kinds of people, and different kinds of economic practices. Yet, my inquiry into pluralism went far beyond these issues.

For example, my observations utilized various sources and media: photographs and written descriptions of sites, photocopies of biologists' observation notes, scientists' illustrations and drawings, videotaped formal interviews with nature reserve rangers, emails with zoo staff, meeting notes with governmental personnel, unrecorded conversations with pastoral nomads, and a range of material sourced from other entities. One of my photographs shows a government archive comprising twelve boxes of documents that contained letters, receipts, grant proposals, newspaper clippings, advertisements for technologies, and more. From the vantage point of pluralism, it was obvious that these items did not cohere as one single thing and that they could not be brought under one unitary framework because they were so dissimilar. It seemed obvious that the very idea of reintroduction existing as one single thing was a fallacy, a misunderstanding arising from talking about the abstract concept of "reintroduction biology." In fact, this concept did not exist independently of my use of it to name a network of plural entities. In fact, the term is an analytical description that makes possible a particular kind of perception. I began to enquire more deeply into the significance of theories of pluralism, a kind of metapluralism, and how my experience of "reintroduction biology" related to them.

METAPLURALISM, ANTHROPOLOGICAL THEORY, AND INTERNATIONAL REINTRODUCTION BIOLOGY

Pluralism's inherent fragmentation raises questions of metatheory, provokes discussions about the potential for theoretical generalization, and addresses the problems and possibilities of drawing broader inferences from anthropological observations. A metatheoretical reflection on pluralism itself, a concept that seems to be universal across all objects and events, holds that there can never be one principle of explanation, not only reintroduction biology, but as any seemingly singular phenomena is more than one. Yet, pluralism is one principle that in this case holds for the explanation of all cases. So, from the perspective of metapluralism, I cannot describe or invoke a unitary concept of pluralism to describe and explain the reintroduction biology of Przewalski's horses. Instead, the pluralisms of pluralism itself became a direct object of my critical reflection. Therefore, this book is as much about the concept of pluralism as it is about Przewalski's horses or their reintroduction biology.

The first form of pluralism that I discuss is ontological. Ontology is the study of what is there metaphysically, beyond what is simply apparent to our senses, and is concerned with what the qualities of existing things are. In this book, I explore the ways that scientists' descriptions of horses' behavior differ between three disparate contexts: a captive horse population, a released horse population in a semi-reserve, and populations fully reintroduced to large wild areas in Mongolia. When first thinking about the horses, scientists assumed they were unitary objects with particular fixed qualities. Through their experiences with the horses, scientists increasingly thought of them in terms of ontological pluralism, the idea that the horses themselves were plural entities with variable qualities. Although she did not use the term "ontological pluralism," one scientist named Maria Noddack came to argue that every individual horse's behavior varied physically, emotionally, and attitudinally, and what is more, this behavior changed through time. My reflections led me to believe that all entities in the world that we perceive and think to be unitary with fixed qualities are also ontologically plural.

The second form of pluralism I discuss is epistemological and concerns the sources and conditions of knowledge about the horses. Due to the complex nature of the horses and their different environments, scientists encounter many difficulties in observing and monitoring them. I describe the plural ways of knowing, the different ways that human individuals interact with the external world of the horses, and the ideas formulated by a range of observers. A multiplicity of methods has contributed to the knowledge of horses and their behavior: these include indirect, technologically-mediated methods, direct observation and the use of intuition and imagination, and a non-intentional, being-in-the-world with-the-horses. Thus, in addition to ontological pluralism, there is epistemological pluralism which draws on and mobilizes many sources and conditions of knowledge. In the context of an ontologically plural world, it is logical that knowledge of such a world would require many sources and conditions.

Thirdly, pluralism refers to the many different languages used by scientists to think about and communicate their perceptions. Languages are not just different groups of sounds and grammars, but vary at a semantic level, at the level of meaning. There are a number of languages used in reintroduction and this linguistic pluralism relates to the variety of forms of reasoning and thinking about the horses and their behavior among scientists. For example, written descriptions of horse behavior can be quite variable; some biologists have researched "aggressive" behaviors and others have researchedsurrender behaviors as constructive of dominance heirarchies, rendering the species either more competitive or more cooperative. As I learned, the languages and vocabularies of scientists are tools or technologies which, in drawing on structurally embedded cultural assumptions, cut apart the ontologically plural world of the horses. They achieve this by guiding a scientist's observations and expressions. Pluralism, in this sense, does not simply refer to multiple material, physical worlds or the ways that we might come to know about that world, but also to the varied means by which we find meaning about the world and communicate between ourselves, as humans, about that world.

Pluralism also refers to the variety of cultural values that form the basis of human participation in reintroduction. For example, when considering the monitoring of released horses in particular, individuals, including both scientists and nonscientists, use different criteria and standards to design and implement monitoring systems, and to participate more generally in reintroduction. It is important to remember that such criteria and standards are essentially cultural values and ideologies, normative conceptions of an ideal state of social affairs. The integration of, and conflict between, the different values of cultural groups into international reintroduction biology is clear in discussions the have about which monitoring system is best. I believe there is a logical necessity to such breadth of opinion as different cultural groups use different forms of language, thought, and reasoning within the context of different forms of social organization to make sense of a complex and plural material world. This kind of disparity is indicative of a complex set of human objectives that include both abstract knowledge of the world, and as an intersubjective endeavor, the attempt to realize and improve human societies.

Finally, pluralism refers to the broader state of mind and body associated with particular sentiments that occur in reintroduction. While science might entail observation and thought, it is also concerned with how people feel toward things as much as how things are observed, how they are perceived, and how they are thought about. Whereas scientists use a variety of methods and languages, their practice is also part of a popular sentimental pluralism, a range of feelings that promote, undergird, and oppose activities like reintroduction. Nonscientists may happily see reintroduction as advantageous for community development, or quite angrily see it as reflective of processes of bias and corruption, even as harmful to their livelihood. It is the very range of sentiments which is present in the social interaction between scientists and nonscientists that makes reintroduction possible—and compelling.

STRUCTURE OF THE BOOK

I explore these philosophical arguments about pluralism in reintroduction biology through what has been called "narrative ethnography" (Geertz 1988; McClusky 2001), in which a story informed by ethnographic data is used to make a broader social theoretical or philosophical point. Anthropologists' engagement with narrative ethnography is a product of the problematization and reflection upon older varieties of scientific realism and writing in anthropology. Many anthropologists now see all their representations, regardless of their scientificity, as necessarily incomplete and perspectival (Marcus and Fischer 1986). This awareness is part of the general critical reflection on scientific practice more generally, and scientific writing in particular. It has led to new ways of exploring anthropological argumentation and to narrative ethnography.

I forward my argument about pluralism in Przewalski horse reintroduction not only through theorization and conceptualization, but also via several interrelated, but independent narratives. As I analyzed my field notes and data from the research process, I chose, like the anthropologist Laura McClusky, to "include certain things, in certain ways, and . . . exclude other things altogether. Information in {this narrative ethnography} is therefore not raw data, but is rather carefully chosen and presented" (2001, 15). I present a narrative, not a scientific argument. It is based on my experience, field notes, and research data, but these are combined in ways that attempt to convey what I ultimately learned, not just what I saw or heard, as a story. The text consists of six chapters that explore pluralism within the processes and practices of reintroduction biology in contexts ranging from zoo research to community development initiatives around horse release sites. Finally, I provide an analytical conclusion. I have changed the names of participating individuals and some institutions to protect the confidentiality and privacy of those concerned.

In chapter 1, I consider a small zoo enclosure in Germany's Cologne Zoo, a long-time home to captive-bred Przewalski's horses, established well before scientists began reintroducing the horses to the wild. I address the changing knowledge of German biologists as they (and others) worked at converting the material and environmental circumstances of some of these horses from captive to wild conditions. During the 1980s biologists at the Cologne Zoo began to use science to research the social behavior of these horses so as to improve their captive management. I introduce the work of Wendelin Schroeder, a biologist involved with the reintroduction of Przewalski's horses, and explore the descriptions she and her colleagues Frieda Franck and Maria Noddack formulated about the horses' behaviors as they dealt respectively with a captive population in Germany, a released population in a Hungarian semi-reserve, and finally, populations reintroduced to large wild areas in Mongolia. I utilize the story of these scientists and their changing observation of the horses to explore my argument about ontological pluralism.

In chapter 2, I discuss Wendelin's involvement with a reintroduction in China. With the assistance of the Cologne Zoo and another partner institution in the United States, the Wild Horse Breeding Center in China developed a captive breeding program for the horses in the late 1980s. Both the German and the Chinese scientists hoped to make observations of the horses' behavior after release to the wild. Over a period of twenty years and with the assistance of German institutions, Chinese personnel built up a herd of over 100 animals. They then began a release and monitoring program in the nearby Kalamaili Nature Reserve (KNR). I discuss the story of an American biologist, Kurt Austerlitz, a Han Chinese nature reserve ranger, Qiang Feng, and an ethnic minority Kazakh local resident, Muhamet Kajat, as they each struggled to monitor the horses, to explore my argument about epistemological pluralism.

In chapter 3, I describe Chinese reintroduction biology and the distinctive research of Chinese scientists at the Kalamaili reintroduction site. After an unsuccessful first release attempt, the Chinese Wild Horse Breeding Center prepared a second release by moving 40 horses from captivity to an acclimation enclosure in Kalamaili. Within the large enclosure, the horses were allowed time to adjust to the reserve conditions before release. While the enclosure gates were open during the day, the horses were fed supplemental hay inside the enclosure—a practice which encouraged most of the horse groups to remain in the general area, returning to the enclosure for feeding at particular times. This acclimation enclosure set-up greatly facilitated short-term behavioral observations. After time spent studying horses in the enclosure, a group of scientists and students from Beijing Forestry University wrote up their observations in the Chinese language. I tell the story of how Dr. Yang and his students, Wang Jie and Jiang Ximing, noticed otherwise unremarked behaviors of the horses and offered distinctive ways of explaining these behaviors, using Chinese-language semantic concepts, to explore my argument about linguistic pluralism.

In chapter 4, I describe how Wendelin and I attempted to develop a formal participatory monitoring project in which local nonscientists, specifically, ethnic minority Kazakh individuals and their families who live around the reintroduction site, would be incorporated into the project as rangers with responsibility for monitoring released horses. I discuss how, in conjunction with the monitoring work already being done by reserve staff, quadrants near reintroduction sites would be assigned to a family to patrol and record information concerning horse groups. Kazakh people belong to cultural and linguistic minorities in China and Mongolia, and describe horse behavior in their Turkic language, potentially utilizing ways of knowing and semantic concepts that might offer new insights into the horses and their behavior. The Kazakh are pastoralists who move among seasonal pastures with their tents and livestock: some of their pastures are located in the Kalamaili Nature Reserve. The aim was that they would prepare reports—with additional translation and interpretation assistance provided by the project—that would allow their observations to be incorporated into databases shared with other rangers, management staff, and scientists like Wendelin. This chapter explores the story of this project and how conflict developed between project personnel, including the Chinese director Li Wenjun and one of the local Kazakh residents, Kadai Akenbek, who was recruited for the project in an attempt to establish participatory monitoring. Through this story, I explore my argument for value pluralism.

In chapter 5, I discuss both the range of sentiments and the interests of the nonscientists in Mongolia who have been affected by the reintroduction of the horses to the area that has become Hustai Nuruu National Park in Mongolia, a participatory monitoring project was successfully established. I discuss the experiences and views of Munkjargal, a Mongolian woman, who passionately opposed the reintroduction of the Przewalski's horses. I document her pain and sadness, and record the various responses of other reintroduction supporters and skeptics, Batyn, Dash, and Demchigiin. Despite a successful implementation of a participatory monitoring program, there were conflicts—not just at the level of values, but also at the level of nonscientists' sentiments towards the project. Using the story, I move from an examination of values in science to an exploration of the diverse sentiments that motivate scientific knowledge and institutions.

The concluding chapter ends with a discussion of the importance of pluralism in understanding international reintroduction biology. While reintroduction is made possible by the scientists' important work and sincere dedication, those scientists often search, urgently, for single best practices to overcome problems; in so doing they try to develop a consensus on knowledge, formulate standard monitoring programs, and evolve community development programs. But the broader and unifying argument of this book is simple; pluralism in both nature and culture suggests that any single "best practice" will prove incapable of resolving the fundamental problems related to the reintroduction of Przewalski's horses, or, for that matter, those of virtually any reintroduction program. I suggest instead, that resolution of these problems will involve squarely recognizing the pluralism inherent in all observations, in monitoring systems, and in reintroduction programs—an approach likely to be more beneficial for science itself and for the communities affected by its practice.

I applaud scientists' efforts to mitigate biodiversity loss and improve the natural and social worlds. But, my conclusions show that there are and will be many uncertainties concerning the future of the horses, and people, despite reintroduction and development efforts, and they still face many kinds of natural and social problems. There are a number of issues that make a basic positive assessment of reintroduction questionable. There is a lot of reflection needed on future reintroductions, the changing historical and environmental context in which species reintroduction takes place, and changing priorities of scientists and the communities they work with. For an indefinitely long time, reintroduction will face conflicts and undergo transformation.

But to begin, I turn my attention to the horses themselves.

ONE
Horses and Ontological Pluralism

My discussion of pluralism in science begins with the problem of a single reality of the horses. The relationship of scientists with the horses, as with human-animal relations in the work of other anthropologists (Descola 2013; Kohn 2013; Viveiros de Castro 2014), leads me to make ontological, nonsymbolic claims about reality. Pluralism in the sphere of scientific culture, as I now understand it, exists because of the presence of a much more basic pluralism in nature (see Scott's [2007] related concept of "poly-ontology," or multiple, different ways of being). While sciences like biology are, in part, attempts to address environmental problems and make society healthier, safer, and more sustainable, they are also attempts by individuals and individual minds to develop a coherent and comprehensive understanding of the physical complexity of nature. For example, scientists are interested in why the horses behave in the way they do as well as in the ways that external and internal stimuli prompt particular behaviors. While helping them understand the horses, the study of horse behavior also expands scientists' understanding of the behavior of the world, of the universe. But that world cannot be fully circumscribed in understanding by any individual scientist for ontological reasons. Through reflection on and comparison of scientists' divergent observations, I believe certain dynamics and properties of the horses' reality are revealed in a profound way.

In discussing the scientific understanding of the horses themselves, the animals at the center of reintroduction, I begin in Cologne, a large city on the Rhine River in western Germany. There, I met a scientist who has made in-depth observations of the Przewalski's horses living in the Cologne Zoo. The zoo is one of the oldest in the world and a leader in the European Endangered Species Program, a management tool for species like the Przewalski's horse. By discussing the horses with Cologne zookeepers who hold biology doctorates and who have conducted research on Przewalksi's horses, I first began to understand the complexity of these animals and what I call their ontological pluralism—the idea that their essential being has multiple modes which transcend any single set of statements or knowledge. This realization was part of the process of the scientific discovery made by the scientists themselves as they made new observations that challenged existing understandings. Their recognition of the horses' pluralism initiated the development of new theories aimed at explaining the horses' behavior.

I met Wendelin Schroeder, a biologist working on Przewalski horse reintroduction, at the main gate to the Cologne Zoo. Despite her age (approximately 60 years old), she looked quite youthful, dressed in a simple blue T-shirt and jeans. We sat down and chatted in her office, with its potted plants, and a picture window that looked out onto the zoo. Behind her desk, there were shelves with many binders containing the zoo's animal breeding records. As we chatted, zoo colleagues came in and out of her office, and she would occasionally check her email on her desktop computer. Everyone was proficient in the English language, although they used German in speaking with one another and they published more often in German than in English. Later that day, we went down to the zoo and walked among the exhibits, chatting as we strolled along the paved walkway that led to the Przewalski horse exhibit itself. Although I came to conduct a formal interview with Wendelin, we already knew one another through a number of previous meetings and trips to a reintroduction site.

In the past, scientists like Wendelin have observed the behavior of the horses within the zoo enclosure and have developed their scientific knowledge about horse behavior through the analysis of such observations. The enclosure she showed me contained a breeding group of the horses (one male and several females), and the zoo she said, kept another single sex group of males (known as a bachelor group) in another enclosure outside the zoo grounds. Here at the zoo enclosure, the breeding group included one stallion and three mares in an area covering approximately 3,000 square meters. The relatively small enclosure was separated from zoo patrons by a small fence, and inside it, there were numerous trees as well as a feeding and watering area.

For Wendelin, "social behavior" is a term which refers not just to the way in which the horses act in relationship to other horses. The term also embraces their response to other aspects of the external environment, as well as expressions of their own physical and internal factors, such as their bodily states and their consciousness. One aspect of the horses' social behavior that particularly interested Wendelin was their agonistic or dominance behavior. Often the horses behaved in ways that were unexpected, but Wendelin thought that their unpredictably was an illusion. She wanted to be able to understand and predict their agonistic behavior and to describe the chain of cause and effect in such behaviors, if only originally, for assistance in management. As Wendelin would discover, scientists face a significant scientific problem when trying to generalize from their observations of individual horses and apply their theories to the behavior of other groups and sites.

Originally she was interested in violent behavior that caused harm. Wendelin defined dominance behavior as an individual horse's access to greater resources gained through competition, aggressive fighting, and the establishment of hierarchy rankings. In her parlance, individual horses compete for access to limited resources and reproduction opportunities, and an expression of dominance indicates the success of a particular individual. Through its dominance behavior, an individual horse aims to establish greater access to resources and reproduction opportunities. Through time, Wendelin thought that the individual horses within a group would develop a fixed dominance hierarchy, a set of unchanging and established ranks for each horse. Rather than fight aggressively for a particular rank each time they interact (which would waste energy and be maladaptive, according to evolutionary paradigms), Wendelin originally believed that the horses would develop stable, ranked relationships. For example, a linear and stable ranked hierarchy is where horse A is dominant over horse B, and A and B are dominant over C, in all interactions. In her view, this dominance hierarchy would allow for less direct competition and aggression within each interaction between two or more horses. She believed that the horses' behavior was systematic and predictable and assumed that similar principles would be shown to be involved in establishing stable dominance hierarchies across different groups of horses and sites of reintroduction. Yet, she discovered a very different kind of phenomena through her research.

In this chapter, I draw upon the twentieth century French philosopher Gilles Deleuze's philosophy of difference and diversity (1994 [1968]). Deleuze's work includes a noteworthy description of the constitutive role of difference in the ontological structure of reality. In this, difference, understood as the characteristics that distinguish one thing from another, contrasts with identity (that which unifies one thing and another (see also Heidegger 1969 and Holbraad 2012). For Deleuze, difference is an essential component of reality. He contends that organisms possess plural attributes, each of which is potentially uncountable: the inherent plurality of such attributes allows them to appear as an infinite number of different things. Deleuze acknowledges those properties which are fully realized as well as those tendencies and capacities which are potential and yet to be realized, what he terms "virtual" (see DeLanda 2002). In this view, there is no single individual horse: each horse is multiple, passing through different states and displaying different behaviors, and each repetition of behavior is different from the one that came before. Difference itself, in Deleuze's view, is what is most fundamental, and is more important than either the diversity of organisms or their properties. As he says, "Difference is not diversity. Diversity is given, but difference is that by which the given is given" (1994 [1968], 202). Rather than existing merely as an abstract idea for Deleuze, the act of reflecting on difference brings us closer to understanding the being of the horses, and it allows us to accept that each scientist will make different observations of different animals, and come to different understandings. Rather than reflecting "error," these plural understandings form the very basis of the nature they seek to understand.

For Deleuze, the existence of discrete entities presupposes difference; it is constitutive of their very being. Difference is what is most fundamental and ontologically basic to reality. At the same time, it undermines the order and unity of any one thing, be it an individual animal or a dominance hierarchy between animals—a precondition which also calls into question the ability of scientists to create universally applicable generalizations. Thus, Deleuze's philosophy is helpful in understanding the problems Wendelin faced when new observations were made which challenged her original assumptions about horses' behaviors as fixed properties and qualities. The application of Deleuze's work raises questions about scientific goals in knowledge production, particularly in relation to attempts to formulate a conclusive description of the social behavior of an individual species. How could a single description be applicable in all times and all places? How might such a description be capable of accommodating multiple observations, anomalous phenomena, or a lack of replication or disagreement amongst observers? The only answer I have is ontological pluralism.

DIFFERENT AT EVERY SITE

The perspective of early singular and unified understandings of horse behavior is problematic. Biologists, in the early period of captive breeding, thought that Przewalski's horses have regular behavioral patterns associated with a general harem-like polygynous social organization in which groups are made up of one male and many females. In this view, the number of females in a group might vary between one and several dozen individuals. Adult males, stallions, live in bachelor groups from the time they leave their natal harem at maturity until they acquire mares of their own. German biologists argue that, despite the presence of certain forms of intra-group cooperation, group members often compete with one another in an attempt to gain access to resources. Competition of this kind is said to lead to the establishment of intra-group dominance hierarchies (Houpt and Boyd 1994). This principle informed the way Wendelin organized her group of male and female horses at the Cologne Zoo.

Although Wendelin was influenced by earlier biological theorizations of behaviorism and was looking for a determinative factor to explain dominance and hierarchy among the horses, she discovered that diverse factors were involved in the formation of different dominance relationships. She did not use the Deleuzean concept of difference, nor did she use the term ontological pluralism, but she noted it nonetheless. For example, when Wendelin first began her research, the Cologne herd included a stallion named Bogdan and three mares named Olha, Vasylna, and Yeva. Wendelin noted that any changes to the membership of the herd led to changes not only in the behavior of the individuals but also to alterations in their hierarchies. For example, the three-year-old stallion Bogdan had the lowest rank in the group at the beginning of her research, but as he aged, and two new mares named Masha and Liuba were introduced in the following years, Bogdan's behavior and rank changed; he began to direct agonistic behavior towards Liuba, and rose in rank above both of the new mares. So, Wendelin thought that age was a contributing factor as well as the presence of particular individuals in the formation of different dominance relationships.

She discovered that there were numerous other factors that played a part in these particular horses' behavior. After a serious fight between them, Bogdan rose above Yeva, one of the original mares, and became second to Olha and Vasylna, the oldest mares. In Wendelin's view, this singular event was determinative in the formation of their relationships. Furthermore, Bogdan's rise in rank caused the group to split into two subgroups of mares, Masha and Liuba, and Olha and Vasylna. Clearly, there were plural factors both within and outside Bogdan himself that led to these changes in the group's dominance hierarchy. When Wendelin looked at each event that had changed the dominance relationships, she saw that the particular factors were different in each case, and the emerging picture was quite complex.

Further observations in the enclosure supported this pluralist idea. Wendelin described how, as Bogdan aged, he began herding and driving the lower ranking mares, a behavior that enabled him to climb in rank. Wendelin said that such herding is an instinctual behavior associated with keeping members of the group concentrated spatially within the wild. Although "natural," this aggressiveness within captivity nevertheless had a negative effect on the mares, a situation which required Wendelin's intervention to ensure the mares' better access to food and to reduce the stress they were experiencing. As she separated the animals, she noted how the behavior of the mares changed again as they split into new groupings. She said that the horses showed different behavioral states at different points in their lives, and she began to understand that there were multiple factors in their behavior and hierarchies, and that dominance was always fluid and shifting. I am explicitly concerned about Wendelin's understanding here, as it reveals a multicausality, where different factors can interact, forming new entities with causal agency, through time.

During 1985–1986, Wendelin made methodological observations which both supplemented and complicated her earlier records and verified the diverse range of factors impacting on dominance behaviors. Talking about differences between the two batches of data, she said,

> When you work with the horses for a short time, you see a deceptively simple hierarchy of dominance and submission. But, this gives you only a superficial understanding. When you study them for a long time, you will find constant changes in the hierarchy. This was completely different than what I had originally thought. In fact, hierarchies are not stable and change through time.

This is more than simply Wendelin's decision to engage and enact a rigorous science. Her comments show how she realized her previous theoretical oversimplification. For example, Wendelin said that the group hierarchy changed after she returned the stallion Bogdan to the enclosure. Firstly she noted the disbanding of the new group that had formed when the stallion was first removed. Then she saw how the two mares Olha and Vasylna began to struggle against one another to achieve a higher ranking. Wendelin said,

> The balanced relationship between Olha and Vaslyna changed when a fight took place between them. Olha, who won the fight, attacked Vaslyna so violently that she did not come near her anymore. It was quite sudden and difficult to explain.

Whereas the mares Ohla and Vasylna had formed an attachment relationship in the absence of the stallion, they were now agonistic towards one another in his presence. This was not a simple event repeated again and again, but complex and occurring once and only once. In my review of the literature, these kinds of idiosyncratic observations were not a standard part of behavioralism in biology and mainstream ideas of dominance hierarchy, although biologists often informally talked about them. Yet, following Deleuze, I argue that the existence of different entities presupposes this kind of difference; it is constitutive of their very being, an ontological pluralism.

After four months Wendelin stopped her daily observations, and she began to reflect on her data and its complexity. Chatting with me in one of the zoo's conference rooms, she reflected back on the period,

> Naturally, I was aware of the fact that social behavior varies considerably among horses and that unusual, distinct behavior and the individual character of one animal may be decisive for relationships within the group. But I don't think I realized how significant singular events are for the animals, even if this isn't the case in a mathematical-statistical sense. Such unusual behaviors and singular events can be determinative of phenomena such as dominance hierarchies.

Wendelin was part of the second generation of zoo personnel who were referred to as "scientists" and who replaced a different kind of collector/zookeeper. Despite a century of captive breeding, there wasn't an extensive literature on the horses, even less so one informed by a modern researcher like Wendelin. She had been part of the horse breeding program for more than twenty years and during that time she had experienced, and observed, several iterations of their groupings. Wendelin discussed the importance of single and rare *events*—moments of difference—because she knew the difficulty of conceptualizing species behavior with determinative principles when change happens from event to event, moment to moment.

Wendelin added a handwritten note onto an early typed research document, a draft of an article she later published. This piece was about how the possibilities for social behavior are likely to differ from zoo to zoo in a way that reflects behavior as a shifting set of many different singular states (and systems). By singular, I refer to individual animals as well as to any traumatizing event that might change the animal's subsequent responses. Each behavioral state's emergence will relate to the horses' internal biological cycles and mechanisms, to other horses, and to the external environment at a particular moment of time. She wrote,

> In this sense, the hierarchy of the Cologne herd is extraordinarily complex. Clearly, the network resulting from changing hierarchical relationships is so complex that it is extremely difficult to describe. Relationships between individuals are not stable, and hierarchies differ in each particular group of Przewalski horses due to the great number of possible factors. That means that agonistic behavior, relationships, and hierarchies will be different at different sites.

There is nothing that can be thought of as a fixed "identity" here; rather, it is a relational ontology in which relations constitute multiple differences and beings. Animals are born, grow, mature, and change over time: each animal's life course is unique and so there are simply no identical social relationships which reoccur through time. Gradually, as scientists interacted with and observed the horses, they began to realize this phenomenon and considered how it might inform their science.

The ideas of the eighteenth century philosopher and mathematician Gottfried Wilhelm Leibniz, who influenced Deleuze (1993 [1988]), are useful for thinking through Wendelin's idea of number—"a great number of possible factors." Leibniz was interested in basic metaphysical questions such as "what exists in the world?" and "what are the world's basic substances that constitute the world?" In his *Monadology*, Liebniz (1965 [1714]) describes a world that is made up of an infinite number of distinct entities, metaphysical points. Everything in the world, for Leibniz, was made up of these points, which he called monads. He describes a world made up of an infinite number of monads, each different from one another, and themselves made up of an infinity, in terms of number, of smaller monads. The monads were mathematically infinite, enveloping a series of multiple sets. Each monad has different and changing qualities; Leibniz says that this is precisely why the world is constantly changing.

Leibniz's monads help us to think through Wendelin's observations of the Przewalski's horses: an analogy can be made between monads and horses. Following Leibniz, we might say the horses are sets that envelop multiplicities. As Leibniz says, "The infinite multitude of simple substances creates the appearances of as many different universes" (1965, 157). We might say that different horses *at different moments of time* create the appearance of infinite horses capable of expressing an infinite range of behaviors. The interaction of these infinite monads, both internally within an entity, and externally between them, lead to a potentially infinite expression of behaviors. Leibniz allows us to imagine each horse as being made up of so many different factors that the prediction of behavior or hierarchical rank would be impossible. His idea provides a way to begin to think of dominance hierarchies among individuals and groups of horses as more complex than any single principle would be capable of describing. The broader significance is, of course, that this conceptualization can prospectively be applied to any object that scientists might observe; it is not simply about Wendelin, this particular scientist, nor is it just about the Przewalski horse, this particular object. This, I have termed, is ontological pluralism.

If scientists are interested in developing a coherent and comprehensive understanding of horse behavior, such ontological pluralism raises problems for generalization. If generalization is the process whereby conclusions are drawn from one population and then applied to the behavior of a whole species, Wendelin's findings would suggest that there is no uniformity across the dominance and rank behavior of the individuals in a group and therefore no general conclusions about the formation of such hierarchies can be drawn.

UNFORESEEABLE COMPLEXITY

In my second example of ontological pluralism, I continue to multiply scientists' descriptions of the complexity of horse behavior. After Wendelin's original project and the publication of her findings in the 1980s, the Cologne Zoo began work on reintroduction of the horses to semi-reserves—large protected land areas considered suitable for the release of the horses in France, the Netherlands, and Hungary. In one such project, begun in the late 1990s, scientists released horses into Hortobágyi National Park, an 800 square kilometer nature reserve in eastern Hungary. The region is made up of the largest plains and wetlands in Europe, and has been used by herders for grazing animals over thousands of years. The park provided a 24 square kilometer fenced area known locally as Pentezug; it was bordered by a small river and channel on three sides, and fencing was constructed to demarcate the fourth side. The site was larger and generally quite different from the horse accommodation at the Cologne Zoo. Here a different context for different horses—a semi-reserve rather than a small zoo enclosure—introduced additional factors that allowed for a much extended range of behaviors and possible dominance relations.

As the reintroduction began, Wendelin sent a young PhD candidate named Frieda Franck to Hortobágyi. The Cologne Zoo cooperated with the University of Cologne, and when I met Frieda she had recently finished her PhD in biology. For her dissertation research in Hungary, organized by Wendelin, she was observing the horses for the first time outside zoo enclosures. We talked, and she told me about her behavioral observations of the Hungarian horses. Frieda made observations for three years during 1998, 1999, and 2000, and she said it was difficult to describe a hierarchy for either of the project's two groups of horses in simple terms. Like Wendelin, the focus of Frieda's research was on agonistic behavior and dominance hierarchies, but Frieda's study concentrated on unsettled dominance relationships: these involved interactions between horses that could either not be observed at all, or not adequately observed, or those in which calculations ended up being inconclusive and therefore incapable of showing any one horse to be truly dominant.

In her research in late 1999, one group of horses was known, after its stallion, as Markus, and was made up of four mares named Jlona, Beatrix, Marta, and Catus. They, like other groups of horses, would exhibit a range of behaviors that included feeding, resting, and dominance, but the frequency with which different behaviors were exhibited was different than among the groups that Wendelin observed at the Cologne Zoo. Frieda said that the percentage of time that the animals spent doing various activities differed because of the greater amount of space available to the horses. She explained that the total number of social interactions decreased because the horses simply had more space and thus had fewer opportunities to interact. For example, during the bulk of the time that Frieda observed the horses as they were engaged in grazing, they were generally less stressed and less aggressive. In her opinion there was an inverse relation between time spent in feeding and time spent in dominance behavior. So, in addition to the role of the age of an individual and the presence of and relationship between particular individuals, the space and activities of the horses were also determinative of a group's dominance hierarchy through time.

At the same time, Frieda struggled to establish any dominance hierarchy for Markus' group. For his group, she could not establish a linear ranking where Horse A was dominant over B, and B was dominant over C, and C was dominant over D. Unlike stallion dominance observed in captivity, the mares Marta and Beatrix were dominant over the stallion Markus, although a rank order between Marta and Beatrix could not be clearly established. In describing these relationships, Frieda made comparisons with Wendelin's descriptions of the horses' dominance behavior in captivity, and suggested that linear hierarchies were far too simple a means to categorize the behavior she had observed in Hungary. She said,

> An important way that I tried to show this was by describing triangular relationships in which A is dominant over B, B is dominant over C, but C is also dominant over A. In a group based on another stallion, I noticed there were triangles between the mares Ursula, Soffya, and the stallion Ervin as well as between Ervin, Soffya, and Sara. Then, I began to note them more often. This made for difficulties in trying to adequately describe their dominance relationships. To say the least, it was confusing.

Frieda also said she documented polygonal relationships in which horse B is dominant over A and D, but subdominant over E and F. The semi-reserve environment and a different, larger group had clearly led to very different behavioral outcomes and much more complex hierarchies. She said she was not sure how to explain these behaviors and the range of possible factors that might be determinative.

Thus, Frieda noted the complex set of other factors involved in producing social relationships and hierarchies among the horses—for example, body condition, parenthood, emigration, immigration, and duration of membership in a group. She continued,

> The factors that determine the rank of a horse are diverse. There are probably even idiosyncratic factors such as an individual horse's temper and experience which have a role. In any one context, the importance of a particular factor can be different and thus make rank formation extremely difficult to understand. At this point, there are simply too many factors to begin to calculate.

Other biologists have noted similar phenomena. In Franck's view, horses and their behavior are not just things, but are elements of a whole complex ecosystem that include the biotic and abiotic components of the total environment. She noted, for example, the use of the biotic components of the environment by horse groups in Hortobágyi and its influence on dominance behavior. She noted the natural grassland of the Hortobágyi puszta, and the distinctive short and salt grasses found there. She thought that some behaviors might even be related to the ways that the horses interacted with each of these distinctive plant species. These factors are, for most intents and purposes, like Leibniz's monads, infinite or innumerable.

Following the eighteenth century empiricist philosopher David Hume's theories (1963 [1748]), the material reality of the horses is the source of Frieda's optical sensations, her observations, and her analysis. That material reality is a force whose vigor and vitality exceeds what either Frieda's sensations or observations, or her analysis can capture in the words of her reports. Thus, when Frieda says there are multiple factors, we can hypothesize the existence of an innumerable quantity of factors behind her insight. The analysis produced by Frieda's observations would be, in Hume's view, as deficient as memory is in relation to the lived actuality of a historical event. It is faded, deadened, lacking in precision, vague and general in relation to causal material reality itself. When she perceives multiple factors, Frieda's attempt to find repetitions and patterns in her observations is independent of empirical reality. Our recognition of identities and patterns is simply the product of this faded and imprecise, vague and general *memory* about reality. Thus, Deleuze (1991 [1953]) drew upon Hume in developing a discussion of the importance of understanding subjectivity for understanding the empirical knowledge we discover and the general patterns that we think we find there.

TO MEASURE AND GRASP

In my third example of ontological pluralism, I show how the increasingly large and open contexts of reintroduction are associated with further behavioral changes among the horses. Wendelin and the Cologne Zoo were involved with international reintroduction with the hopes of both increasing the probability of establishing a wild population as well as creating new opportunities to observe horse behavior in the wild. In 2001, the Cologne Zoo embarked upon an expedition through supposed Przewalski horse habitat in Eurasia, crossing from Hungary to Ukraine, Russia, Kazakhstan, China, and Mongolia. Reintroduction, in the views of these biologists, is not simply release outside zoo enclosures; the term implies that the reintroduction will be exclusively effected into an area where the reintroduced species lived originally. During the expedition, partnerships were formed between the Cologne Zoo and various universities and nature reserves, and as a result, reintroduction of Przewalski's horses in each of these countries mentioned is ongoing. Wendelin and her colleagues in the international biology community have begun to conduct studies of released Przewalski horse behavior at each site and Mongolia has become very important in the continuing study of horse behavior.

A number of studies have been conducted in Mongolia's Hustai Nuruu National Park. As the scientific observations and conclusions concerning dominance hierarchies at Hustai challenged earlier assumptions, research conducted at this site presents another opportunity to think about not only the observable ontological differences and diversities across the horses but also that which cannot be known about them. The Przewalski's horses were reintroduced to Hustai in the early 1990s. At 500 square kilometers, Hustai is a much larger site in comparison with the 24 square kilometer horse reserve in Hungary. Hustai's particular environment hosts various distinctive plants and other biotic components. There are far more horses present in Hustai than any one scientist could monitor since many breeding groups use mutually exclusive territories which are separated by mountain ridges.

Maria Noddack, a colleague of Wendelin's, studied the horses' dominance behavior in Hustai. She was in her thirties and had a passion for the horses and the Mongolian region. In her dissertation and numerous published articles, she developed the concept of nonlinear dominance hierarchies. Her observations at Hustai differed significantly from those of the scientists who had researched horse behavior in Hungary and Germany, and rather than attempting to resolve the issue of complexity,

Maria's work contributed to an increasing academic recognition of the value and interest inherent in the horses behavioral complexity.

In Hustai, Maria tried to establish an understanding of dominance hierarchies for four groups named after their respective stallions—Chulun, Akhbattar, Temur, and Batukhan. Chulun had four mares named Altantsetseg, Erdene, Tuyaa, and Oyunbileg as well as two other mares that joined the group at a later time. Each mare had given birth to foals. After some years Chulun eventually lost his mares to Batukhan. Ankhbattar, the second stallion, had five mares named Bayarmaa, Khongordzol, Tuul, Oyuun, and Jochi. His mares gave birth to few foals, and many died of an infectious disease. Finally, Temur had seven mares named Enkh, Monkh, Naran, Ajarchan, Arghun, Dojoodorj, and Yisun. His group had a number of mare movements in and out of the group, and a significant number of births. In general, these groups were significantly larger than the ones in Germany or Hungary, and Maria considered group size to be a factor that affected social behavior.

Like Wendelin, Maria originally felt that linear dominance hierarchy was functionally important because it provided a way for horses to develop a social order that decreased agonism and competition for resources. In a linear dominance hierarchy, each horse had a fixed position, and therefore, did not have to continually struggle with other horses in the group for resources. In questioning the findings of Wendelin and Frieda, Maria tried to be more rigorous in both her data collection and her analysis so as to establish linear hierarchies in the midst of complexity. For example, she only documented agonistic interactions in which a clear winner was able to "move" another animal, rather than simply recording any and every agonistic action and eliminating any relationship in which a dominance hierarchy could not be established. Moreover, she used a number of sophisticated mathematical-statistical tools for analysis (see De Vries 1995; De Vries and Appleby 2000). As the size and number of breeding groups increased, Maria used these tools to look for patterns in a large and complex data set of observations, a process that mirrored the usual strategies of wildlife biologists, rather than those of zoo biologists.

In the context of complexity and uncertainty, mathematical data can seem like a foolproof means to capture certainty (Bogost 2015). Maria used proportional analysis to establish which horse won the most encounters, that is, proportions of those encounters won versus those attempted. Using the I and SI method, an iterative algorithm, Maria analyzed the data of those relationships between animals without a known linear rank order—for example those occupying various ranks in a triangular or polygonal hierarchy. Rather than allowing these complex and unclear relations to skew her analysis, Maria used these statistical tools as a means to simplify and thus deal better with her findings. Her aim was to reduce the "noise of complexity" of the horses, and to achieve a simplification or probability, and thus, an approximate understanding vis-à-vis a linear hierarchy. Traditionally biologists have investigated entities that are located in real space and time, but when such objects are studied using mathematics they are transformed into abstract entities. Although scientists may be motivated to understand the actual animals, and their mathematical data might be thought of as related directly to their concrete empirical experience, it is analytically discernible from that experience.

Maria concluded that the linearity of dominance hierarchies varied amongst individuals and individual breeding groups as well as across periods of time, and she pictured this as a curving and varying line. She thought that nonlinear aspects of the relationships existed and that these reflected two issues: firstly, the prior attributes of individual animals which help enable dominance but which change through time such as height, weight, and aggressiveness. Secondly, she took other important dynamics related to individuals' relationships into account; these include the accumulated effects of winning or losing prior competitions, the effects of bystanders on the agonistic interaction, and the length of tenure within the group of the two particular individuals. Maria argued that dominance hierarchies cannot be predicted based on the prior attributes of individuals and such hierarchies have to be understood through the concrete interactions of the individuals in real time. In her view, hierarchies develop via sustained interactions through time, and linear hierarchies would require a greater repetition of interactions than were seen in the Hustai groups which were so frequently changing. Moreover, she also added that it was very rare for the horses in Hustai to have limitations in access to particular resources that are only available to one horse at a time (such as in captivity). Therefore a linear hierarchy might not necessarily be adaptive in a truly wild context. Even with the simplification afforded by mathematical tools, Maria was left with ontological pluralism.

In sum, Maria argued that dominance relationships among horses are best described as self-organizing and self-structuring, and they are effectively unstable and nonlinear. Maria told me that she began to reject the linearity of hierarchies—the idea that each two horses have a simple additive or multiplicative relation to each other. In contrast, nonlinear relations between horses, for Maria, could not be expressed by such simple calculations, and their mathematics were likely to involve subtraction and exponential relations. In this way, they can be complex and even chaotic, in that differences in each of the factors discussed, small changes over time, and all the feedbacks among these factors can easily lead to divergent outcomes for hierarchies. Like other nonlinear systems, the shapes these hierarchies take are likely to be unpredictable.

CONCLUSION

This chapter has explored how scientists attempted to describe individual horse behaviors, both the personal behaviors of individual horses as well as those which are applicable to individuals across the species. What was ultimately discovered was that each description of individual horse behavior was like a singular perspective on different individual horses and horse groups. This is the ontological difference that exists between our concepts of particular kinds of beings like Przewalski's horses and individual entities such as this or that horse that we perceive.

In one sense, the ideal abstraction of whole species behavior begins to break down with each observed sample of horse behavior. By questioning the whole species model it becomes possible to realize a state of diverse, unique beings, single specimens, and pure difference. At the same time and more philosophically, I think that there is no single individual horse; each horse is multiple, passing through different states and displaying different behaviors, and each seemingly repetitive exhibition of behavior is in fact quite different than the one that came before.

Pluralism holds that there can never be one single principle in explanation and yet ontological pluralism is the one principle that, in this case, provides an adequate explanation for all the diversity that we have seen. Rather than invoking ontological analysis as an explanatory principle for divergent descriptions of horses, ontological pluralism itself is really the object of my critical reflection. As Deleuze might say, difference itself is what is most fundamental; it takes precedence over either the diversity of humans, other biological organisms, or any of their seemingly infinite properties. This holds true for my own description in relationship to that reality I describe.

In the next chapter, I explore the scientific search for observational data of the horses within the context of a reintroduction. The chapter will consider reintroduction's open and porous relationship to various methods, and importantly, its relationship to various people, including scientists, in China. I turn from ontological pluralism to an examination of difference seen through the lens of epistemological pluralism.

TWO

Epistemologies in Monitoring Released Horses

The production of scientific knowledge about Przewalski's horses, in addition to the horses' ontology inself, entails a very complex social, cultural, and epistemological system. Because the physical reality of the horses is so dynamic and unpredictable, scientists have spent a lot of time trying to perfect their observational methodology. These observations are very important. Through their observations of the behavior of the released horses, scientists have come to better understand the species' individual and group behavior and their knowledge has contributed to a successful reintroduction strategy. Rather than simply reflecting human curiosity and a will to know, scientists' observations include this applied conservation component. Much discussion about observational methodology takes place within this practical problem solving of the day-to-day management of reintroduced horses.

Beginning in the 1980s, staff and personnel from the Wild Horse Breeding Center in China developed a captive breeding population of Przewalski's horses, and by the year 2000, they had built up a herd of over 100 animals. In 2001, the center started its first reintroduction program in the vast Kalamaili Nature Reserve, a part of the Dzungarian Gobi in the north of the Xinjiang Uighur Autonomous Region, China's most northwestern region. Interestingly, the center's monitoring strategy did not resemble monitoring systems used at other reintroduction sites, such as Hustai Nuruu in Mongolia. At first, 30 horses were released, and the horses began a phase of exploration, acting unpredictably. Reserve rangers had difficulty locating them. Following a more successful second release, a new herd of horses utilized a large, unfenced natural area of the Kalamaili reserve, an area the size of Rhode Island (a significantly larger space than in the previously described reintroductions in Hungary and Mongolia). By 2006, there were some 40 horses freely roaming in the reserve. To assist rangers in monitoring, international biologists brought new observational technologies to Kalamaili such as indirect satellite monitoring. Whereas international and Chinese scientists were interested in a range of behavioral issues, the reserve staff wanted to try to describe basic events such as births, dispersals of individuals from groups, and the condition of injured and ill individuals as well as group locations, movements, and home ranges of released horses. As with the behavioral science described in the previous chapter, staff members used direct visual observation, supplemented with technologies like binoculars. Officially, the reserve relied on a ranger-based monitoring system; the Chinese project had a dozen personnel and four vehicles, but lacked other equipment that is routinely utilized in much international-grade wildlife monitoring—for example, surveillance aircraft that make ranger-based observation easier and more effective than exclusively ground-based operations. Regardless, there was no single method that was central to how the horses were monitored in Kalamaili.

Rather than being a method involving an individual person using a sense perception to understand a world of objects, monitoring of the horses actually entails complex social interaction across a diversity of people. As I have shown, knowledge of the horses is never given by any one person in its totality, but instead, different people know different things about the horses; through social relationships, these individuals share their knowledge, and learn from one another. In their monitoring system, staff integrated the observational methods used by professional scientists, trained and experienced rangers, and local pastoral nomadic people who lived around reintroduction sites. Differences exist in what is known about the horses across these observers, but each observer also uses different ways of knowing them too. Different people observed the horses in relation to different sources and conditions of knowledge. The synergy between scientists, rangers', and local people's knowledge highlights the epistemological pluralism that is required in making basic animal observations in the context of ontological complexity.

In his model of human epistemology, sociologist Pierre Bourdieu (1977) elaborated a theory of practice in which he, like Deleuze, rejected the traditional division between objective and subjective knowledge. This division, lodged at the center of scientific method, is based upon the idea of knowledge as an intentional and abstract representation of a thing developed mentally through sense impressions. For example, scientists argue that one should look out onto the horses using visual sense perception to develop a mental image and knowledge of the horses. Knowledge, in Bourdieu's view, despite one's method, can never be such a mirror-like objective representation of the world, in which human knowledge depends wholly upon the nature of the world itself and an unbiased and undistorted human sense perception of that world. Furthermore, a purely subjective knowledge, whereby representations of objects only emerge from the mental conditions of thought, totally separate and distinct from the world itself, never exists. For Bourdieu, this binary understanding of knowledge, and how it creates dichotomies between more or less objective knowledge, is highly flawed.

Bourdieu sees knowledge as the product of a complex dialectical relation between the things in the world, the objective, ontological conditions of knowledge, and the subject, the human knower and his or her distinct conditions of knowing. Knowledge is diverse because it reflects a particular practice of an individual in the world, different positions in that world, different experiences, habits, and life. It is situational. Bourdieu recognizes that different methods—different ways of being in the world—can allow for different kinds of qualitative knowledge about that world. Drawing on Bourdieu and some of the philosophers who influenced him, I describe and explain the epistemological pluralism of scientists, rangers, and other people involved in monitoring in Kalamaili as distinct knowledge-producing practices, and show how such a variety of practices—different technologies, methods, and even ways of life—were valuable to the monitoring of the horses.

SATELLITE AND COMPUTER-MEDIATED INDIRECT OBSERVATION

In the global context of international reintroduction biology, there are a number of North American and European scientists who now assist Chinese staff in monitoring Przewalski's horses (Leimgruber et al. 2005; Romeijn et al. 2003; Walzer and Kaczensky 2008). One of these scientists, Kurt Austerlitz, works at a large U.S.-based conservation and research center and focuses on remote sensing and computer applications in conservation biology. He is an English and German-speaking biologist who trained in undergraduate biology in Germany, and carried out his PhD in the U.S. When I spoke with him he was clearly passionate about conservation and science and highly enthusiastic about utilizing remote sensing technology to assist the Przewalski horse reintroduction. His remote sensing technology allows the project to obtain information about horses from a distance—some 6,000 miles—without the need for direct observation. In remote sensing technologies, a navigational and mapping system such as the Global Positioning System (GPS) uses satellite technology to determine an object's Earth position and this is used to track the horses and their movements from a distance. Kurt told me that remote sensing is useful in a context like Kalamaili where, due to monitoring system limitations and environmental conditions, direct observation is difficult. His use of remote sensing is my first example of a distinctive epistemology.

Practices such as remote sensing have many implications for understanding epistemological practices in Przewalski horse monitoring. As part of the range of possible ways of gathering and generating knowledge, remote sensing technologies like GPS also involve particular individuals, each with their own distinctive skills and techniques, machines and equipment, and their own interaction with a set of ontological conditions. Rather than simply producing an objective knowledge about the horses, the work of scientists like Kurt reveals certain aspects of the horses, whilst concealing others, and inasmuch as it reduces some uncertainties, it also introduces others. Although Kurt has overcome some aspects of the difficulties of direct observation, a diverse range of natural and social problems, both natural and social, still remain. While remote sensing is limited in its ability to solve all aspects of monitoring problems, Kurt's work has nevertheless revealed new and distinctive aspects of the behavior of the horses.

I met Kurt on a number of occasions in China, and then subsequently, traveled to his research center in Virginia too. He met me at the airport in a white cargo van, and as we drove, he described his research center. His workplace was a large, public facility on several hundred acres of land surrounded with security fencing. Most of the buildings were simple cinderblock structures. The site was home to important work on breeding captive populations of endangered species and it included laboratories devoted to remote sensing of habitats, and other conservation-related research. Like the scientists at the Cologne Zoo, Kurt was involved with a number of international conservation projects as well as the Przewalski horse reintroductions.

To allow for improved observations in Kalamaili, Chinese staff invited Kurt to participate in the project. In 2006, he arrived in Kalamaili as part of an international team composed of a biologist, a veterinarian, and a technician, each of whom was accompanied by Chinese counterparts. Using dart guns, the team tranquilized and collared several Przewalski's horses. These collars were fitted with GPS and a VHF radio device to allow the team to remotely sense the horses' location by reception of microwave radio signals transmitted from the collars. In 2007, several more mares were collared. In discussing this technology with me, Kurt said

> A collar attached to the horse transmits a signal to an orbiting satellite at programmed times. We then receive the signal from the satellite to our computers. The collars record up to four GPS positions of the collared animal—recorded in longitude and latitude—in one day. After we record it, the locational data for the horses is then brought into a GIS program, and we can monitor the horses' movements through time visually. After collaring the animals on site, the monitoring can then be done from here at our laboratory.

I looked around at the computer terminals, and the desks laden with satellite collars and coffee cups. There were scientific posters on the walls showing elephant conservation in Myanmar, and around us interns were busy at work. Kurt continued,

> From a laboratory like this, I can provide assistance to rangers in locating the horses, or any other wildlife too. Biologists have been using remote sensing for a long time; today, desktop computers with GIS can easily utilize locational information and biologists can layer information on topographic and vegetation maps. It's really widespread. In fact, GIS has become a very important conservation tool.

While he provided basic locational information to staff, Kurt also hoped to be able to explore the horses' behavior on large temporal and spatial scales. He was interested in researching the horses' habitat selection, and the behavior of the animals in selecting areas of habitat to live. He wanted to better understand how the horses were adapting to the local habitat and its resources. He continued,

> We are combining our horse location data with satellite estimates of biomass productivity around the release site, and trying to predict how

> the horses will move in relation to biomass. Biomass is the weight of living organisms such as vegetation. This data is acquired by imaging technology on board NASA's Terra satellite. The satellite's sensors image the entire earth every couple of days, and thus produce a lot of data about an area's vegetation and its changes. We layer the horses' locational data with these vegetation data sets in ARCGIS, a specialized program for data visualization.
>
> Originally, the idea was that the horses move towards areas with high biomass. From our research, we saw that they can actually move away from such areas. As we now understand, high quality forage resources often have lower biomass in summer, and rather than vegetation, water seems to become a more determinative factor in their habitat choice. Our research has shown our original hypothesis to be incorrect, and has opened up new ways of thinking about Przewalski horse habitat selection and behavior.

While not capable of locating the horses with absolute certainty, Kurt's findings clearly improved the project scientists' understanding of the behavior of the horses after reintroduction. His approach simplified, and in a way distorted the complexity of the horses as it captured some aspects of behavior and allowed others to remain quite unknowable.

As I continued to talk with Kurt, he also spoke about many problems with the satellite monitoring. For example, there were issues with faulty batteries on the satellite collars and other technical problems; collars accidentally fell off animals and sometimes others were lost when their wearers died. In a long process involving tranquilizers, dart guns, and Chinese customs, he had to return to Kalamaili numerous times to re-collar horses. Most importantly, by the time the locational information was sent from the collar to the satellite, then by email to Virginia, then forwarded to the regional administrative office in Urumqi, and finally sent by cellphone to Qiakeertu and then the rangers, the horses had often changed locations. As much as they wanted to be able to "see" in real time as in direct monitoring, there was often a technological and temporal *lag* (Boellstorff 2008, 101). Therefore, this method of monitoring the horses was not always useful for producing real time locational data for staff. Other problems also illustrated how the equipment interacted with the material conditions of the horses to create uncertainty and thus required other methods and ways of knowing them.

For example, the collars were also equipped with transmitters which functioned in the normal radio wave spectrum. Radio telemetry offers staff another form of indirect monitoring; they can use small portable transmitters attached to the animal's satellite collars. The radio waves emitted by these can be picked up by staff through a handheld antennae. Sometimes, the receiver would help point them in the right direction, but in this region, radio telemetry was often beset with difficulties. The canyons and crevices of the Kalamaili reserve often blocked the signals, much as a radio fades out when you are traveling in an automobile and pass through a tunnel. In general, the locational data was incomplete and of uneven quality. Kurt said,

> A lot of different things can happen with this technology out there. It's not a magic bullet. We definitely need to continue to collar animals. But satellite monitoring is just supplementary. Staff still need a range of other methods in the practical management of the reintroduction.

As I listened to him, I realized that Kurt recognized the limits of his methods and knowledge. With his comment about the "magic bullet," he made reference to the Freischutz, a German story about a hunter who, by a contract with the devil, had obtained a number of bullets destined to hit his prey without fail. Unlike the marksman in the story, Kurt recognized he would never have a supernatural knowledge given in its totality.

From this example, I have shown how understanding an epistemology in use entails understanding the particular physical locality of the horses, the characteristics of the equipment an observer uses, and their general practice. Pierre Bourdieu (1997) argued that we should both recognize the limits to all knowledge, and realize that any one set of practices is incapable of giving us infallible access to a complex world. In the next section, I discuss the unexpected practices I discovered that are used by rangers to observe the horses. As you will see, the ontologically complex world might even require what the seventeenth century physicist, mathematician, and philosopher Blaise Pascal (1966 [1670]), called the "heart." By the heart, Pascal meant what can be intuitively known as opposed to what can be known through empirical observation.

KNOWING THEM FROM THE INSIDE—FROM THE HEART

From Virginia, I made my way to China to talk with rangers about their monitoring methods on the ground. From the airport in Beijing, the Kalamaili reserve is a long, three day, meandering train ride across the vast Chinese landscape and along the course of the Yellow River. The train follows the river's course to western China's Tibetan Plateau, and then follows the plateau's northeastern edge until it arrives at the arid Xinjiang region and the Tianshan Mountains, where it heads northwest. The train journey ends at the regional capital of Urumqi, a major urban area with a population of almost three million people, a portion of whom are Muslim and Turkic-speaking Chinese ethnic minorities such as Uighur, Hui, and Kazakh. My destination was a further four-hour drive north of Urumqi; I passed along the main highway, leaving the city and travelled across the vast expanse of Kalamaili, past oil and coal fields, and the small regional towns along the way.

After crossing the reserve's border, I arrived at Qiakeertu, a small town and nature reserve headquarters. I made my way to the Kalamaili Conservation Station, a three story government facility painted light blue with white trim. This institution serves as the headquarters for rangers that observe and monitor reintroduced Przewalski's horses. As I entered the building, the office and dormitories were vacant or closed, the only person around being a man named Qiang Feng. He was short and stocky, and wore inexpensive and practical army clothing, a colorful sweater, and a grey jacket with passants, or straps, along the shoulder. He was from central China, spoke Mandarin Chinese with the influence of a local dialect, and had a master's degree in wildlife management. He sat working at a desk. From that day and over the course of the next several months, Qiang and I would often sit together and talk either in his office, the monitoring station, or in the reserve truck.

Qiang showed me the landscape of the Kalamaili Nature Reserve, a beautiful desert grassland and shrubland with a low mountain range. The reserve also included several small marshes and oases, an area of the Guerbantungut Desert with large sand dunes, a dry and rocky plateau-like grassland, and a series of small rocky crevices with seasonal streambeds. The horse release sites were located in the grassland, and in 2006, a group of the horses were present in a border area of the plateau, with sand dunes on its southern border and a large canyon (with fixed and seasonal water sources), some 100 miles long. While the rangers knew that the horses were in that canyon, it was a complex topography in which to locate, let alone observe, them. To understand how they monitored the horses, Qiang said I had to get to know and understand this landscape.

With Qiang's permission, I spent time with him while he monitored horses in the reserve. I would stand with him in Kalamaili as he scanned the landscape and look out on the desert grassland. As he had the most experience of any of the rangers, he had an assurance about him and his work that made our interactions natural. Other rangers thought foreign researchers were critical of them; foreign biologists had criticized the reintroduction and said that the rangers did not use scientific monitoring methods. Indeed, Kurt had sent staff to train them in observational techniques, and as Qiang told me, there was conflict between his staff and at least one of the rangers. "We have our own way of doing things here," said Qiang.

Rangers like Qiang had accumulated many hours, days, weeks, and years of experience in monitoring, and such people were invaluable to the basic logistics of horse monitoring in the reserve setting. In terms of its day-to-day operation, the Chinese reintroduction project really depended on rangers like Qiang, even if the knowledge they had was not always perfect and was itself filled with uncertainties. Despite their problems, the monitoring reports rangers wrote are very significant in the everyday management of the horses. Translated into English, these reports were also an important information source for foreign scientists like Kurt. At the same time, the rangers' experience of monitoring released horses was limited to the years following the first reintroduction in 2001, and they often had problems in simply locating the horses. Much of their experience had been obtained at the breeding center in work with its captive population of horses, rather than during monitoring reintroduced populations in the landscape of the Kalamaili reserve. In Kalamaili, locating the horses was locating something small in a much larger area.

The epistemological practices of Qiang and his team of rangers include distinctive equipment, ways of knowing, and work practices that had varying degrees of success. Each team member had their own distinctive skills and techniques, and while not perfect, these did enable people to locate the horses with some frequency. From reserve headquarters, four rangers monitored the horses, and worked in two teams. They switched on and off and rotated every few weeks. When they were not monitoring, they returned to Qiakeertu to write reports and finish other paperwork. Each two man team divided the monitoring work up informally along with the tasks of taking care of the horse acclimation enclosures and maintaining monitoring stations in the reserve. They interspersed repairing their vehicle, a Toyota truck, carrying water to the stations, and other duties, with taking care of and monitoring the horses. Both teams had to overcome certain monitoring problems and to accept the limitations of some of their methods—just as Kurt did.

On one of these spring monitoring days, Qiang and I were sitting in his truck and talking. As we watched several groups of horses off in the distance, Qiang said,

> Winter is the easiest time for monitoring, but when spring comes, the horses move over great distances, and then we have to track them throughout the reserve. But, when we have to track them, we are lucky to get a basic visual contact before they flee from us. They are wild, and they have a very large flight distance; they simply do not let you get close to them. Under these conditions, we can't make detailed observations. We look for secondary evidence of their presence, and maybe a short visual contact and count. As summer approaches, they shift from using seasonal water sources to using fixed, permanent ones. Then once again, they are easier to find.

The concrete place in which observation is done and the human body's position in it has consequences for how scientific observation is carried out (see Goodwin 1995). To summarize, Qiang did not have difficulty locating the horses seasonally when they were located around fixed water sources, but had difficulty when the horses moved with the emergence of impermanent, but widely available, seasonal water sources. His observational methodology also varied accordingly. As I would learn,

Qiang's method of observation during this more difficult period was not just empirical, but also intuitive: he used instinctive and imaginative feeling, and not just conscious perception and reasoning, in relation to this particular place and his body within that place.

That said, Qiang did make use of empirical evidence. Rangers like him used what American hunters call "spoor"—the horses' tracks and paths, their hoof prints, and feces—as well as other signs of movement and activity. Like the American hunters, rangers used these evidences to infer the animals' subsequent activities and to attempt to locate them. Since a hunter rarely begins with visual contact of his prey, he or she must use indirect or secondary signs and make inferences from the signs to the movement, direction, and thus the location of wildlife. These signs are evidentiary, but their use is logical, based on the memory and past experience of the hunter or observer. They are most definitely synthesized with intuition and with various other elements of experience and the world. Qiang said,

> If you have good fresh tracks, you can easily tell what direction they are going in. If there is fresh dung, you will know that they are nearby. If I see fresh dung, I might spend hours right there, scanning. Climb up onto the top of the truck, and use binoculars to scan the area every few minutes. You just know. More commonly, we try to visit water sources they are using. They need to drink water once or twice a day, and so they can always be found within ten kilometers or so of a water source. Then, they are easier to find. If you get into their mind, sometimes you just know where they are. You think they should be thirsty and so they are going to go to a nearby spring.

Struggling to explain what he did, he continued,

> Sometimes, I try to find an area of higher elevation for observation. This makes scanning more productive as I have a larger field of vision. But sometimes, the horses are behind a landscape feature and you just can't see them. Then, you have to be patient and wait. They might move out from behind the feature and you will see them. Other times you don't; with early morning or afternoon light, the horses can be lost in the shadows of landscape features. You think you see a horse, but in reality, its an onager, a kind of wild ass. Sometimes I see something, but I am not sure what I am seeing. You trust your instincts, and sometimes, you are right. Sometimes you are wrong.

As he later said, he was not always able to use empirical evidence. When pressed about his method of observation during the most difficult times, he said he literally "became like a horse." Since he could not always physically locate them, his goal was to learn to understand the motivation of the horses' behavior and anticipate their action using an imagined point of view. In Qiang's monitoring, to observe the horses in the wild depends upon imagining what the horses are thinking or willing, inferring what they are going to do and where they are going to go next. Qiang used this tactic to provide one way of knowing reality—a way that includes both unobserved and unobservable elements. He contrasted this with how, as he said, he saw foreign scientists tranquilize and collar the horses physically for satellite monitoring; "They come to know them from the outside," said Qiang, "but we try to know them from the inside." This is what he meant when he said he "became like a horse."

Reintroduction biology can be said to actively require this use of intuition and imagination, as does science more generally (Holton 2000). As shown by the example of Qiang's method, the real, ontologically plural world presents problems and limitations in utilizing one particular method, and other epistemological practices are needed to deal with the problems at hand in different settings of science. In the situation described above, Qiang has access neither to adequate remote sensing data, nor to data produced by direct observation. In the context of reintroduction and monitoring work, rangers like Qiang are forced to act and to act quickly; they cannot always see or indirectly perceive the horses before having to act. In practice, their expertise might entail what others have called "unconscious and undeliberated embodiment" (Dreyfus 1987). Pascal (1966 [1670]) and Bourdieu (1997) referred to this as "the heart"—an unconscious and undeliberated intuition—and they proposed that it was as important as the senses and mind in knowing the world.

As I continued to talk with Qiang, he spoke about many of the problems they were facing. While his work diminished some of the uncertainty about the horses, he acknowledged that his knowledge was also imperfect. Like Kurt and scientists from the Cologne Zoo, he could not have complete knowledge of the horses. Like Kurt, he had no false sense of confidence in how consistent and reliable his methods were. He knew that he did not have all the answers to pressing questions about the horses, and was aware that the answers he did have were not exact or precise. Thus, Qiang accepted imperfection, inaccuracy, unpredictability, and uncertainty, and this allowed him to be open to other ways of knowing the horses. Like Pascal, Qiang recognized that the being of the world is dynamic and complex, ontologically plural, and thus, our basic epistemological nature is uncertain. "Whenever we think we have a fixed point to which we can cling and make fast, {the world} shifts and leaves us behind; if we follow it, it eludes our grasp, slips away, and flees eternally before us, nothing stands still" (1966 [1670], 92). In many ways, he shared the wisdom of Kurt Austerlitz, and saw the basic necessity of recognizing an epistemological pluralism.

BEING-IN-THE-WORLD: "I HAVE SEEN MOST OF THESE PLACES"

What is the difference between the knowledge of a biologist and a ranger? Kurt might know the horses' movement away from high biomass areas, but only Qiang knows how to track them using spoor and intuition. Together, their knowledge reflects the plural states of conscious human experience. Sometimes these states attend perceptively to the outside world. At other times, they are operating more cerebrally. Human perception moves dialectically between both states in practice, combining them in novel arrangements, and so progresses towards distinctive understandings of the world.

When Qiang had difficulty locating the horses, he consulted nonscientists, ordinary people with no formal role in Przewalski horse reintroduction, but who attend perceptively to the outside world. Even these peoples' experience had a place in constructing observational knowledge about the horses. For example, after having difficulty in locating a particular group of horses, we stopped the truck at an encampment of local herders. In Kalamaili during late fall, winter, and early spring, ethnic minority Kazakhs use the area as pasturage for their livestock. When I met Muhamet, a local Kazakh resident, he was there herding sheep. Even though it was early spring, and the day was sunny and warm, he wore a green *ushank*, a kind of sheepskin-lined hat common in northern China, Mongolia, and Russia. Of course, he spent most of his day, from sun-up to sun-down, outside taking care of his livestock, and so he was prepared for the temperature changes. We talked as he leaned on a walking stick and watched his sheep grazing around us. I later met with him in his yurt, a tent used by the herders, and then several months later, in his home village on the Ulungu River. I learned about how he informally assisted rangers in locating the horses.

From listening to his experience, I learned about how the environment contributed to the physical difficulty of locating and observing the horses. In 2007, the horses moved westward into an area called Maligou, a large valley with several water sources, meagre but adequate vegetation, and a sparse human population. Unfortunately, the area was remote, and not easily accessible by 4x4 vehicle. It was approximately 100 kilometers west from Qiakeertu, and the only access point was by a dirt road that follows the western side of the reserve. As one follows the road south for about two hours, the dirt road drops down into Maligou, a crescent-shaped valley with very rugged terrain. There are numerous drainages and small valleys that criss-cross Maligou, and with its intricate topography and lack of infrastructure, it is very difficult for the rangers to locate and monitor the horses there. Visibility is greatly reduced; there are many areas for the horses to seek shade and shelter and conceal themselves in the landscape. Tall saxoul shrubs further hindered visibility.

Yet, the distinct practices of herders allow for them to locate the horses. The valley has several herding encampments of Kazakh pastoral nomadic people and a county animal husbandry station built to serve them. Kazakhs have used this area for centuries. Local people like Muhamet, who travel off road by horse or camel, know the landscape intimately. They notice changes in the environment, for example the presence or absence of wildlife, and their way of knowing is our third example of a distinctive epistemology in monitoring. When herders using local pastures noticed the horses, word of mouth quickly reached the local animal husbandry station, and station staff contacted the ranger station by CB radio at Qiakeertu. This is fairly common; at particular times, the animal husbandry station will even receive requests from the reintroduction project and ask herders to look for the horses. After surveying and scanning their pastures, herders would come forward and inform the station about any horse movements and sightings (Jiang 2005).

In fact, their informal testimony had become an important part of monitoring. Monitoring in reintroduction biology, like international scientific practice generally, depends on communication by scientists, on testimony. It depends on the social relations between diverse groups that make up its organization. Scientists rely on the testimony of peers, just as the public depends upon the testimony of scientists—the outcome or output of their research (Gerken 2015). But what is less well recognized is how scientists are dependent on the testimony of nonscientists as an input into the process of scientific research. Usually, we regard such lay testimonies as epistemologically distinct from science, but this conceptual distinction is more related to power and recognition within scientific communities rather than their roles and value in the production of scientific knowledge. Local rangers take seriously the world of local herders who, like themselves, perceive the horses through the use of all their senses for both direct and indirect locating. For rangers, knowing the location of the horses sometimes involved this testimony, or using the words of other people who had themselves perceived the horses through sensate means (Coady 1992; Foley 2001). The use of such knowledge depended on the rangers trusting the epistemological abilities and the accounts of local herders and then using those accounts to guide further observation.

At first glance, these skills and techniques would seem to have nothing to do with the kind of intentional observation seen in Kurt and Qiang's practice. They are rooted in the subsistence strategy and way of life called herding. In response to a question about his assistance finding the horses, Muhamet said, in his distinctive herder manner,

> About the only thing that I truly know is animals and this pasture. I've ridden a horse, and I have seen most of the places around here. Yes, my eyes can see and I can speak. I have told the rangers about where I have seen the horses, and they have asked me to help on three or four occasions. Last time, there was too much snow, and the rangers had trouble with their vehicle. In winter, it's better to travel in Kalamaili by horseback. While previously herding, I came upon the horses grazing on my pasture, and knowing where they were, I took the rangers to the site. It really is quite simple.

In considering the epistemology of herders like Muhamet concerning the location of horses, there is a question of how they find the horses, their practical method. Obviously, they see them. But how they see them—when others look for them and cannot see them—involves a distinctive practice, their total way of life in Kalamaili and its skills and techniques.

Herding is the central practice that made Muhamet's non-intentional observation of the horses possible. Kazakhs take a month migrating southward from their fall pastures north of Kalamagai and the Ulungu River. In 2009, Muhamet had made nine stops before arriving in Kalamaili, from his summer pasture in the Altai Mountains, not too far from the Mongolian border. Maligou is known to Muhamet as Chemushbai, his winter pasture. That year, he migrated down to Chemushbai in November with three other people—Yerajip, Besen, and Yilgerbek—and they all shared a small yurt, or nomadic tent, on their pasture. The pasture is about 20 square kilometers, and they know this pasture well through their long-term experience of herding.

What is interesting is that they do not intentionally direct their observational skills towards the horses, unless they are asked, and yet their observations are still very useful for science. When I first met Muhamet, he was herding about 300 head of sheep. He had walked out from the yurt for several kilometers, and was beginning to circle back. It was on one of these walks that he had first seen the horses. Herders like Muhamet note animals entering or leaving the area, although he does not formally count or record them. When herding, he would scan the horizon, and might notice the horses (and other kinds of wildlife, including predators, like wolves). This ability to locate the horses is based on the ingrained habit of closely scrutinizing the landscape during the activity of herding.

The thought of Martin Heidegger (1962; 1927), a German philosopher in the twentieth century, is at the heart of the practice theory outlined by Pierre Bourdieu, and it helps us to understand why the knowledge of herders might be essential to rangers and scientists. For Heidegger, scientists and nonscientists are not simply subjects with consciousness that become aware of objects in the world. Both subjects and objects come about through human being-in-the-world, what Bourdieu called "practice." For example, the conservation biologist, the ranger, and the herder are "made" through their practical interaction with the world and the horses specifically by different ways of being and the practices that go with these. Knowledge, in each case, results from the active and dialectical engagement of a human with the immediate world around the human, neither an abstract human individual (and less a scientist) nor an equally abstract world of discrete objects. Heidegger describes this epistemological condition as "being-in-the-world." Our being-in-the-world allows us a practical engagement with the world through our goals and purposes, and knowledge, rather than being the product of a method, is simply the result of these socially and culturally diverse practices. As I have shown in this example, international scientific practice depends upon it.

Science began as a noninstitutionalized practice of so-called amateurs who only seemed to decline as the modern institutions of science overshadowed them. With the rise of institutionalized science, practices such as herding have been devalued in the production of human knowledge, overshadowed by satellites and binoculars. The philosopher Pierre Macheray wisely reminds us, "The first hammer by a blacksmith could not have been a real hammer, any more than the man who made it a real blacksmith, but it was a stone collected from the side of the road, a "natural" instrument, in itself imperfect, which became an instrument only through the use that was made of it, in serving like a tool, which it certainly was not to begin with" (2011, 48). Likewise, the vision of ordinary people thus became scientific observation, clearly seen here as scientists and rangers still need ordinary peoples' observations in Przewalski horse reintroduction biology.

CONCLUSION

A science such as reintroduction biology is usually associated with particular observational and experimental methods. These methods of observation are usually seen as constituting a detachable way of observation, free from concrete goals and purposes, a formal observational methodology. Yet, it is possible to see various forms of lived experience, oriented towards many concrete problems, in particular places, as essential to the production of knowledge about released Przewalski horses. Reintroduction biologists recognize that their capacities are limited in monitoring of released horses and draw upon diverse forms of knowledge and local experience, including that of nonscientists. In the real time of the reintroduction, scientists remain open to data and information that comes from a variety of cultural sources, a great variety of practices adapted to endlessly changing natural and social conditions. This openness reaches into what Helga Nowotny and others (2001) have called science's "epistemological core."

Epistemological pluralism holds for knowing any one phenomena in an ontologically complex world. Because different people bring different things to monitoring, they see the same horses in different ways. In the next chapter, I explore the distinctive Chinese linguistic framework for perceiving, describing, and explaining horse behavior that is at once social, cognitive, and scientific. From the divisions of the horses and how they are known, I want to show you a new pluralism, a pluralism of words.

THREE

The Languages of Observation

The Case of Chinese Science

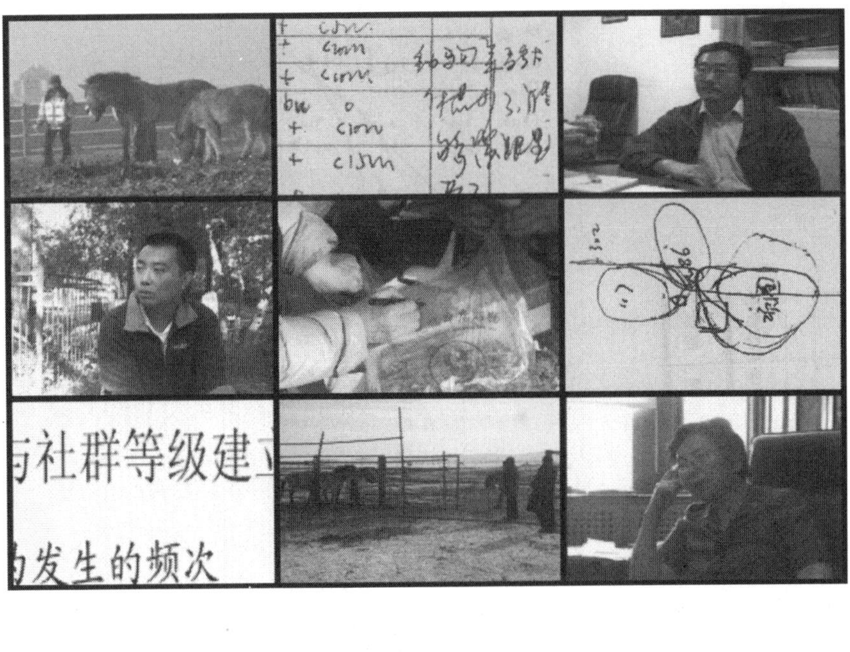

One of the reasons humans form communities is to share cultural practices and develop collective learning. The epistemological practices of nomadic pastoralists, rangers, and biologists are good examples of their respective communities' collective learning. Each develops a "shared repertoire," to use the educational theorist Etienne Wenger's (2000) term, a set of technologies and equipment, sensibilities and routines, and *language* that reflects that learning. For nomadic pastoralists, collective learning is transmitted to one another through a particular variety of the Kazakh language. Similarly, collective knowledge is passed on through a variety of Mandarin Chinese for rangers, and through varieties of German and English for German and U.S. biologists.

After establishing the significance of difference and diversity in the structure of reality, and explaining how pluralism inheres in the ways by which different groups know those realities, I now move on to the significance of linguistic pluralism, the different languages used by people to speak and write, and generally, to communicate about the horses. My discussion focuses on the language of Chinese scientists specifically. In efforts to understand the social behavior of Przewalski's horses in Kalamaili, the research of Chinese-speaking and writing scientists is an important resource for the international scientific community as it provides distinctive insights into Przewalski horse behavior. This chapter looks at a body of biological research, originally produced in Mandarin Chinese, and considers this against the background of the German research described in chapter 1.

The setting for the Chinese research was a fenced enclosure in the northeast section of the Kalamaili reserve. After a number of "hard" release attempts, the Chinese Wild Horse Breeding Center prepared subsequent and "softer" releases by building an acclimation enclosure in the reserve and then moving groups of horses to the enclosure for a period of acclimation. Using the enclosure, the horses were allowed time to adjust to the reserve's environmental conditions before release, hence the idea of a soft release. For a period of several months, project staff kept the new group of horses in this fenced enclosure and fed them supplemental hay. They opened gates to the enclosure each day to allow the horses to utilize local forage and water sources, but due to the feeding, the horses stayed in the enclosure's general area. When the staff prepared to finally release them, they opened the gates permanently and stopped providing hay. Eventually, the group began to graze over more and more distant pastures. Then, the process was repeated with another group of horses.

With horses being fed and thus moored to a fixed space, the acclimation enclosure allowed Chinese scientists to not only monitor, but also conduct in-depth research on horse social behavior. Soon, on government instructions, a Chinese university sent scientists and students to carry out short term observations at the enclosure. Unlike Wendelin's attempt to manage captive individuals and her study of aggressive behavior, this group was much less concerned with violence. Part of the reason was simply the different material conditions of enclosure. But this group also came with the assumptions of Chinese biology—its collective learning—and those embedded in its language. As I began reading their work in Chinese, I realized that Chinese biologists' research also represented a different understanding of horse behavior. Chinese scientists wrote about the horses' behavior in their native language and its stylized ideographic character system. Using these characters, Chinese scientists had produced a number of fascinating descriptions and interpretations of *cooperative* horse behavior (Chen 2008; Chen et al. 2006; 2008; Huang et al. 2012; Zhang 2009). This was in stark contrast with how Wendelin and other German biologists described and interpretted their behavior. My reading of their research, and my formal and informal discussions with them, inform the following analysis of Chinese research on the social behavior of Przewalski's horses. In particular, this analysis describes and explains the semantic differences that exist between German and Chinese biologists' understanding of horse social behavior.

Philosophy that makes a distinction between a subjective and objective knowledge is useful when one considers the degree to which scientists' abstract reflections on the horses and their attempts to explain their behavior are not just ontologically but culturally and linguistically determined. The philosopher Immanuel Kant (2007 [1781]), for example, was interested in the degree to which human understandings of the world had their origin *a priori*, literally before sense perception. In other words, Kant thought that some portion of the human knowledge of objective things might be said to exist before any subjective perception and observation or any particular methodology. Drawing on empiricists like Hume, he argued that if knowledge only came from the senses, it should be highly contingent and variable, with each sense impression distinct from the other, what we call ontologically plural. If it was so contingent and variable, any one observation would be meaningless, devoid of any pattern or justified belief that it was either one thing or another. The French phenomenologist Maurice Merleau-Ponty (2002 [1945]) actually referred to this true empiricism as the human experience of awe and speechlessness. There would simply be no conceptual thing to describe or interpret. Kant recognized that objects in human knowledge *usually* appear more certain, conceptual, and ordered; humans relate what they perceive to what they already know, identify and acknowledge patterns across their experience of things through time and group them into coherent semantic categories and concepts making them part of their particular paradigm, what he called a schema. This is important because we assume scientific categories and concepts come only from objective experience rather than partially from phenomena such as a schema which exists prior to experience.

In this chapter, I consider how Chinese Przewalski horse biologists make observations to test categories and concepts as well as apply concepts and categories to make particular kinds of observations. Post-Kantian philosophers like Johann Gottfried Herder (2002 [1772]) and Wilhelm von Humboldt (1999 [1836]) have argued that language, in relation to other subjective mechanisms Kant described, is essential for understanding this process of how humans attribute abstract meaning to concrete sensation through a priori schemas embedded in language's semantic systems. In this chapter, I explore three ways that the Chinese language allowed native scientists in Kalamaili to think about and give a distinctive meaning to horse behavior. We begin with Dr. Zhen Yang's concept of "surrender behavior."

CHUFU XINGWEI, OR SURRENDER BEHAVIOR

Sociologists of science like Evelyn Fox Keller (1998, also see 1995, 2002) and Donna Haraway (1989) have noted how Western biologies have tended to privilege a schema of competitive interactions between organisms and neglect a cooperative one. After having read Western literature on Przewalski horse behavior, this is the first thing I noticed after reading the Chinese literature. These sociologists are the more recent amongst a number of thinkers to have made this observation of the ideological nature of biology (see Engels 1940 [1883]; Kropotkin 1972 [1902]; Lysenko 1948). In fact, this was an important object of discussion within Chinese biology. Biology textbooks from the Maoist period often opened with a quote from Karl Marx, or the Chinese communist leader Mao Zedong and critiqued Darwinian ideas about evolution and the "struggle for existence," a term which was understood as the competition between individuals for resources. Chinese biology focused on altruistic, harmonious, and cooperative forms of social interaction among animals. In China, the science of biology developed as part of the general revolution against capitalism and the development of a socialist society.

This could be seen in the contemporary biology of Dr. Zhen Yang. Dr. Yang, a biologist in the Department of Life Sciences at Beijing Forestry University, had flown back and forth numerous times from Beijing to Kalamaili, and spent many hours conducting observations at the acclimation enclosure in Kalamaili. I distinctly remember him riding up to his office on an old "Flying Pigeon" bicycle. His office was in a three-story Maoist period campus building which had been recently renovated, and he parked his bicycle at the bike rack by its front door. While he talked, he would frequently push his eyeglasses back up onto the bridge of his nose, eyes wide, and gesture enthusiastically and energetically with his hands. He was a very interesting and intelligent man.

Chinese biologists like Dr. Yang worked in a different social, as well as political and economic, context than their Western counterparts. As he explained to me, Yang provided the central government with advice on policies that require a biological foundation. The central government invited Yang to participate in scientific research in Kalamaili and provided him with research funding. While teaching courses on biology at the university, he also made presentations on his research to China's State Forestry Administration. He now lives in Beijing, but was originally from the Xinjiang region—the area in which the Przewalski horse reintroduction site was located, and the government felt that he both understood the local setting of the reintroduction and could provide a more scientific and objective assessment than was possible in the reports produced by the rangers and administrators in Kalamaili itself. As a result, he and his students went to work on a number of research projects in Kalamaili, and they subsequently published a number of theses and articles based on this research. In terms of the volume of texts written about the horses, Dr. Yang and his colleagues had the largest research program of any group of scientists pertaining to the Przewalski's horses in China.

Chinese biologists like Dr. Yang and his students produced knowledge as well as communicated through the medium of language. Like other scientists, Dr. Yang used an ethogram, a catalog of the species' behaviors accepted by the scientific community, as well as observation charts, to record his research. His ethogram included a distinctive category of horse social behavior called "surrender behavior" (*chufu xingwei*). This term was not utilized by German or American researchers. These two words are noteworthy: they constitute a specialized and technical item of scientific vocabulary. They also form a theoretical concept, subject to a grammatical process in which the verbs "surrender" and "behave" are transformed into a noun phrase, a category of thing, a linguistic phenomena (see Halliday and Martin 1993). The category of behavior that Dr. Yang called surrender was common among individual horses of many different ages; for example, he referred variously to horses walking away, dodging, and running away from other horses. When I first encountered the category, I was taken aback as it presented such a contrast to the behavior predominantly described in English—the aggression of a horse. For Yang, the central subject was not aggression or the aggressor, but the action of what Wendelin Schroeder might have called the *object* of aggression—the *other* individual animal. Yang argued that, in the familial group, social hierarchies based on age create a structure for surrender among lower-ranked horses. He argued that, rather than reflecting dominance and submission, such behavior reflects cooperation and mutually advantageous relationships.

Obviously, for any complex ontology like horse behavior, there are many possibilities for description; I am only using a dichotomy between the horses and knowledge about them as a hypothetical and recognize the empirical relationships that give rise to different schema. But, a scientist necessarily chooses a concept or category from those which are available to him or her such as those that are embedded within his or her existing language and community; the chosen item is most often one used by others and featured in its written literature and verbal discourse. Importantly, when Chinese biologists recognize a horse and its behavior, they choose a concept and category like cooperation that has a specific and significant role in their language. Chinese language, like other languages, provides its own range of abstract categories which are schematic and which can be used to label and understand a variety of sense impressions or observations. Using an ethogram with this category of "surrender," Yang observed horse behavior and I contend that what he observed is not discernible from the language in which he observed it.

In his research, Dr. Yang compared the surrender behavior of horses of different ages as they utilized a water source less than a kilometer away from the acclimation enclosure. He noticed that different horses of different ages behaved differently with each other in relation to water resources; each horse utilized water in quantitatively different ways, in terms of frequency and length of time per visit, and in qualitatively different ways too. Talking about this qualitative usage of water, he said,

> Social animals like Przewalski's horses require functional means to bring each individual into the group. I think age hierarchies provide a basis for group interaction and integration, and so when individuals of different ages interact, they do not interact just in terms of individual objectives, but group objectives too. Age hierarchies provide a position and a function for each individual within the group, and therefore, provide a framework for each individual horse to act. Through the hierarchy, each individual behaves in ways that function for group integration. In my research, I argue that this hierarchy can be seen in surrender behaviors of one horse towards another when accessing water. It allows for a more harmonious group life in which all the horses can gain access to a crucial resource.

This is very different from Wendelin Schroeder's understanding of aggression and individual objectives as described in chapter 1.

I still vividly recall my first impression of Dr. Yang's research. My first impression was that Yang felt much more certain of his conclusions than German biologists like Wendelin Schroeder. The Chinese scholar Min Lin (2001) has written about what he calls the "social production of certainty." Lin argues that certainty does not have to do with the evidential properties of statements, or whether these statements are characterized by an internal logic, but rather, statements have a degree of certainty when the speaker or writer of the statements believes them to be on secure ground, to be justified and validated, and necessary and beyond a reasonable doubt. For Yang, the stability of group relations for social animals like the horses is believed to be on secure ground, justified and validated, and beyond doubt.

Consider how different this understanding is from that of the German biologists described in the first chapter. For example, in Yang's view, a younger horse's surrender to a higher-ranking horse was not an act that reflected the dominance of the higher-ranking horse as supposed by the German researchers, but, instead, a key behavior that helped establish and maintain the group's integrity. In fact, Yang downplayed the concept of aggression, and said that it was only seen when seasonal periods of dryness limited water and fixed water sources brought more than one horse group into contact and conflict resulted over water access. Yang said that stallions, in those particular situations, would only express aggression as their groups of mares came into the proximity of other stallions. He continued,

> Individual members within the same group simply do not compete (*jingzheng*) against one another to get water because members of the same group share the same interests. It is in all of the horses' interests to insure each member of the group has access. One individual horse might surrender to another member of the group to allow them to access water first, but this is not competition or dominance. In Chinese, we do not tend to use the word "compete" to think about how individuals interact socially within a group. Compete, for us, means literally to fight or use violence. I think that surrender among kin is a more accurate rendering of what occurs; it's not fighting, but giving way. It's like a struggle, something that is difficult and requires effort, but it is more functionally effective for the group than aggression. It allows an individual animal to realize his objectives, but without defeating or establishing superiority over another member of the group. It's more harmonious.

In this example, the cooperative social schema embedded in Mandarin Chinese and the language of the biological community clearly provides a resource for Yang to analyze the horses' locomotive patterns, movements, and interactions in ways that differ from those chosen by English and German speaking and writing scientists. This resource is a schema, in the Kantian sense (Kant 2007, 176) by which ideas and understanding become associated with sense perceptions. To borrow and extend Kant, schema allow scientists, building on their different native languages, to analytically break down the otherwise continuous locomotive patterns, movement, and interactions of animals and attribute different conceptual meanings to them. In the next section, I explore one of Dr. Yang's students and his related concept of "gathering behavior."

JUJI XINGWEI, OR GATHERING BEHAVIOR

In this section I consider a second example of Chinese biologists distinctive conceptualization of horse social behavior. In China, there are distinctive philosophies of cause and effect (*yinguolun*); the Chinese philosopher Whalen Lai (1981) has argued that in traditional Chinese philosophy the cause of any one thing does not have one source, and neither are any two things wholly independent of one another. For traditional Chinese philosophy, things are, rather, born of a confluence of conditions. This distinct concept of causality is still highly influential in Chinese thought and is one of the reasons why Chinese biologists describe and explain horse behavior differently than Western ones. In fact, causality was one of the main schema that Kant himself described (2007, 180). I explore this in the research of Jie Wang.

Jie was one of Dr. Yang's brightest graduate students, and he had a particularly interesting understanding of the horses' social behavior. When I first met him, Jie was in his late twenties, and was from a rural area of northern China. He was the first in his family to have attended university, and he talked about his family's poverty. In Beijing, Jie and I met to discuss his research, and he told me how he had researched the "gathering behavior" of the horses. As Jie told me, his research was a fragment of Dr. Yang's larger state research project, and Dr. Yang wanted him to focus on the group relationships seen to be essential to the horses' reproduction and population growth. Whereas violence was an issue for captive horses, the cohesiveness of horse groups was seen as most important for reintroduced horses. In a way that I had not discerned within the German science, Jie described the arrangement of horses within their group in spatial terms. He described how they moved across space synchronously, and he highlighted the changing distance between each individual within the group during activities like grazing. Chinese ideas about cause and effect can be seen in his discussion of group relationships, and in particular, his description and explanation of "gathering behavior."

A schema of cause and effect, in Kant's view, is like images, representations of the external forms of things, although he conceived of an image as being more specific and concrete than a general and universal concept (2007, 178–79; also see Matherne 2015). An image is a simple representation based on a few sense impressions of an object. For example, Jie made a number of observations that led to his formulation of a ribbon-like image of horse movement. In Kant's work, the image is different than a concept, an abstract understanding free from any concrete sense impressions—(for example, a concept like "group relationships"), or what Kant calls an "intuition," a complex empirical representation formed from a number of sense impressions—'the horses are grazing.' The image is situated between the more abstract concept and the more empirical intuition, and thus, is an important functional part of a schema, a middle mechanism by which the mind links abstract concepts of "cause and effect" and "group relationships" with tangible, material sense impressions "horse distribution in space," "these horses are moving." An image is a representation of a particular group of horses, but concepts like "gathering behavior" are a general representation that can apply to all sense impressions of my horses.

With his notion of "gathering behavior," Jie utilized a series of technical vocabulary items that objectified the process of changing family structures in a distinctive way and organized sense impressions quite differently from the ways in which German and other Western scientists interpreted group formation. The Westerners saw group formation as the indirect effect of events which had their cause in individual horse behavior. Individualism, as an important concept in their thinking, allowed them to perceive each individual horse as a causal agent. The terms which Jie and his colleague chose reflected a linguistic emphasis that gave groups of horses, rather than individuals, causal agency. Under the general concept of "gathering behavior," he had created a number of subconcepts, all logically intertwined with this schema. Their relations, taken together with the relations between each vocabulary term, illustrate this schema. In the remainder of this section, I will discuss a number of these concepts and terms.

For example, processes of group change were known as "stallion substitution" (*touma gengti*) and "group division" (*qunti fenlie*). As Chinese scientists usually referred to the horses by their assigned number (part of a sequence of horses, like points that make up a line), Wang described how Stallion #24 had four mares, #59, #76, #36, and #47, and three foals, #124, #136, and #103. The group had formed through stallion substitution (*touma gengti*), with Stallion #24 replacing Stallion #11. Wang described how another group was formed through group division (*qunti fenlie*). Stallion #46 had two mares: #86 and #77 who originally had been part of Stallion #9's group. As stallion #9's group divided, mares from the group left and reduced the size of the original group. Using an ethogram with these terms, Chinese biologists rendered their sense impressions quite differently than their Western counterparts. As Jie spoke, and in his writings, he invoked a clear picture of group shape changing slowly through time. Reading his work, I felt like I had an entire gestalt shift in my perception of the horses' social behavior.

Kant himself influenced the later Gestalt school of psychology. Gestalt psychologists like Max Wertheimer and Wolfgang Kohler tried to understand the human ability to create meaningful perceptions in an ontologically plural world. The central principle of Gestalt psychology, like Kantian philosophy, is that the mind itself forms a whole from this pluralism in a self-organizing manner. Their work in turn influenced philosophers such as Ludwig Wittgenstein (1968 [1953]), who argued that language itself does not simply represent an ordered world, but instead forms a particular kind of order out of a set of human activities and life. He used an example from the Gestalt psychologists of the abstract image of a duck-rabbit, an image that can be perceived as either a duck or a rabbit, but not both at the same time. Looking at the image first as a duck and then as a rabbit is known as a Gestalt shift. For Wittgenstein, the perception of either a duck or a rabbit from this abstract image reflects the ways that humans perceive other things in the world as one thing or another, in relation to human activities and life. Implicit in his formulation is the idea that things, due to their ontological pluralism, can always be perceived as something else. Philosophers of science (Duhem 1991; Hanson 1958; Kuhn 1996) have brought this debate into the analysis of scientific practice with the notion of the "theory-ladenness" of observations, the idea that the theories that a scientist works with will affect the kinds of observations that the scientist subsequently makes. I extend these ideas here by arguing that it is not only the theories that scientists hold that affect their perception, but their general language too. Whereas German scientists might develop theories of aggression and agonistic behavior, Chinese scientists, by contrast, develop theories which emphasize cooperation; both theories draw upon foundational schema inherent in each group's language.

Jie identified other gathering behaviors too: "group combination" (*qunti hebing*), and "individual exchanges among groups" (*jiazu qunjian geti bianhuan*) and he used these terms to objectify and conceptualize behavior in distinctive ways, thus attributing agency to groups far more than to individual horses. After the death of Stallion #18, a new group emerged through group combination (*qunti hebing*): Stallion #17's group grew in size with its combination of mares #64, #49, #67, and #65 with #62, #26, and #60. In another example, Wang described individual exchanges among groups (*jiazu qunjian geti bianhuan*): during a drinking event at the enclosure spring, two groups came into contact without the aggression described above by Dr. Yang. Two mares simply exchanged places in the two groups of Stallion #17 and Stallion #13. Jie clearly used these vocabulary items to objectify and conceptualize behavior in distinctive ways, providing agency not to individual horses within a group, but to the whole groups themselves.

One of the final, and particularly interesting, examples of this schema in Jie's research was "sub-adult secession" (*yacheng liqun*). He said,

> After the young male horses mature, stallions are likely to push them out of the group. This is particularly true if they continue to follow their mothers after a particular age. They have an obligation to leave and form a new group. Mature male horses will search for a new group to join, and produce offspring. They might live for one to two years with other groups, and at around 5 years, they secede from this group

> and find young mature female horses that have left their groups to form a new group. We call this "sub-adult secession" (*yacheng liqun*).

Sub-adult secession reflects a conceptual presupposition that the group is the agent rather than the individual sub-adult horse. In the Chinese view of secession, the behavior of the individual horses simply cannot be described independently of the group; each horse is dependent on the group for their individual being. Thus, any so-called "individual behavior" is described by Chinese researchers as a broader state which applies to a group of individuals seen as a whole. It isn't just that the horses are simply different at each site, or that German and Chinese scientists utilize different epistemologies, but the example of the Chinese Przewalski horse research clearly demonstrates that the different native languages of scientists are an essential part of how distinctive knowledge about the horses is produced. Moreover, reading research in other languages is capable of opening up new understanding for scientists as it can highlight their own conceptual and categorical assumptions, thus potentially improving knowledge too. It is likely that consciousness of assumptions in western research even played a role in the development of this particular Chinese schema.

YINGZI TOUDAO: A NEWLY DEVELOPED CATEGORY

The theories discussed above would suggest that once certain types of objects and processes are conceptualized in language, individual speakers will be likely to notice related phenomena during observations. Or, conversely they may be limited in their noticing of other kinds of potentially observable phenomena. But, this situation does not necessarily entail ready-made, arrested, or fixed thought. Wilhelm von Humboldt (1999 [1836]), an early writer drawing on Kantian ideas, argued that speech and writing are always dialectical products of both the external environment as well as an individual human's interiority, their own senses, feelings, and thoughts. For Humboldt, language is never statically acquired by individuals, but is dialectically influenced by the lived experience of individuals. For Humboldt, language is not a thing, but an action, what he calls an *energeia*. Through this self-created means, the individual interiority of each biologist will respond to the language, build itself into the language, lend it new meaning, and so allow it to reveal new forms of the world. In this sense, one might, even in the context of the discussion of the schema, focus on differences and diversity present amongst Chinese observers, as each individual biologist responds to the language and lends it meaning.

Ximing Jiang was another student of Dr. Yang's whose thought-provoking research contained a linguistically and culturally specific understanding of horse behavior. I was introduced to Ximing just as she was finishing her dissertation. She was one of only two women amongst Dr. Yang's students, highly respected among her male peers at Beijing Forestry University and known as a particularly keen observer of the horses. Dr. Yang as well as his students recommended that I speak to her.

As part of Dr. Yang's project, Ximing conducted a basic descriptive study of "mare-foal interactive behavior" (*yuyou xingwei*) with observations at the enclosure and an ethogram of behavior she hoped to record. Her ethogram included "categories of nursing" (*buru*), "attention to the foal" (*jianshiyouju*), "using sound to attract a distant foal to return" (*huanyou*), and "mutual movement" (*gensu*). While nursing was addressed in the other languages' ethograms, the latter three categories were not. Indeed, mare and foal interactions played almost no role in the Western literatures' discussion of Przewalski horse behavior, yet these were richly reflected in Chinese research outputs and highlighted the cooperative schema.

What is particularly interesting was how Ximing's work on this topic first began during dialogue with Dr. Yang and was then creatively developed by her when she devised her own categories to describe horse behavior. In this way, Ximing's research demonstrated how the scientific use of language and its particular schema can be creative within a semantic domain or set of vocabulary. With this cooperative schema, Ximing drew attention to an element of the world specifically interesting to herself. She observed and then modified Yang's schema, selecting from, elaborating upon, and transforming the existing Chinese ethogram that he first presented to her. Most important for this chapter, her creativity took place within a linguistically available schema.

In one interview, as we talked about her modifications to her original ethogram, Ximing said,

> Mares have a very special relationship to their foals. They have a gentle and soft, tranquil relation to them. They are usually inseparable. The mares accompany the foals while they play, guard and protect them (*shouhu*), and even scratch their itches (*naoyang*). They take them grazing, and they groom their bodies (*shuli*). All of these are important categories of behavior, and I decided to add them to the categories of behavior that I investigated.
>
> For example, one category (*yingzi toudao*) that I developed came from observing Mare #59 with her foal #124. In the late summer, when it is very hot, foals will look for a cooler place to rest. If they sleep under the hot sun, they risk overheating, suffering heat sickness, and experiencing dehydration. Mare #59 would stand in the hot sun and project her shadow (*yingzi toudao*) for the foal. Once I noticed this mare doing this behavior, and then, I noticed it with others too. It was truly selfless behavior. If there was anything that would cause the mare to be concerned for the welfare of her foal, she would use her mouth to rub and

awaken the foal (*zuiceng, jiaoxing*). These are all important kinds of behavior.

In this example, Ximing simply expanded the semantic range of existing words to create new categories of behavior. For example, she referred to the mare's behavior using a Chinese idiom (*shiduzhiai*), an expression meaning "parently love," but based originally and semantically on how a cow will clean her calf. This broader cooperation schemata allowed Ximing a different and unique ability to analytically break down the movements and interactions of the horses, namely the particular movement and interaction of mares and foals. For her, this relation reflected the Chinese understanding of nonaggressive behavior that was not oriented towards individual goals. The behavioral categories that she used reflected the conceptual presupposition that the relationships between individuals in the group are fundamentally nonaggressive and even altruistic. Ximing's research has shown how an individual will act in ways that favor the success of the members of the group, even when it is at the expense of an individual's own survival, for example, by overexposure to the sun.

CONCLUSION

This chapter has shown how scientists from China who utilize the Chinese language are drawn towards making particular kinds of observations of natural phenomena such as horse behavior. The different languages that scientists speak, write, and communicate in provide a key to understanding the contrasting ways in which they understand that behavior—a factor which is just as important as the differences in the ontological pluralism of the horses themselves or the methods by which the horses are known. When they conduct observations, biologists, both German and Chinese, record them by checking off observed behavior in relation to a list of anticipated behavioral categories designed to test a hypothesis; behaviors that are outside this list will usually be ignored. In observation, any one scientist might combine their own visual acuity with the use of a particular visual technology like binoculars as well as paper documents prepared in advance with written categories of observation expressed in a particular language. This language is as important to observation as eyesight and other visual technologies. In fact, it is a kind of visual technology.

As biologists use predetermined vocabulary items, the observers they employ are liable to relate their particular sense impressions to the presupposed, and sometimes limited, behavior categories which are listed on the ethogram. As mentioned in chapter 1, there is a particular set of vocabulary items (functionally related to the evolutionary framework of adaptions to nature) that dominate this ethogram in the West—the set includes both protracted behavioral states such as feeding, drinking, nursing, standing, locomotion, and shorter behavioral events such as communication or aggression. For example, German ethograms had a number of different categories or vocabularies related to their own theories of aggression, agonistic behavior, and individual competition. Notably, Chinese biologists' ethograms and the vocabulary used in them were oriented to a different political history, a different language of biology, and different theories of social behavior.

In China, the vocabulary items used in this strand of equine research reflect a different way of thinking in science that relates directly to the kinds of relationships that pertain between people in Chinese society. In the Kantian tradition, philosophers and anthropologists have sought to demonstrate how the concepts of human understanding partially emerge out of the distinctive cultural nature of the human subject. Using the notion of schema, I have explored how sensations can be transformed into culturally specific and highly abstract descriptions and explanations. That is, when we speak or write about the world we are not simply or objectively referring to the world or our sensual experience of the world, but we are very explicitly conceptualizing that world using a cultural language. Our concepts and the categories that we use, rather than being simple products of our individual observations, are partially the products of this external linguistic system: they are parts of a learned culture. This is why scientists who speak different languages may refer to the same objects in the world, but still conceptualize them quite differently.

Thinking pluralistically, other languages and the users of those languages might well offer different schema and alternative conceptual presuppositions and thus enable novel and more finely nuanced interpretations of the ontologically plural experience of horses. Having understood the differences between the linguistic concepts used by Western and Chinese scientists, I was intrigued and wanted to find out more about how Kazakh herders, who speak an unrelated Turkic language, might describe the social behavior of the horses. The following chapter documents the surprising outcome of that inquiry.

FOUR

Human Values and the Design of Wildlife Monitoring Systems

In the international world of Przewalski horse reintroduction, German biologists did not always understand the research of their Chinese counterparts. There was miscommunication that cannot be explained by recourse to semantic nonequivalence, poor translation, or monolingualism. Words were translated adequately, but the differences of context that each scientist worked in were much more difficult to translate, and sometimes imperceptible to those who were attempting to communicate. It was not simply people with different languages who were trying to communicate, but people with different values.

The meaning of horse behavior or reintroduction itself never existed as an immediate, evident, and uncontestable fact to everyone involved. There was rarely a meaning that corresponded to a particular horse, or a group of horses, or the reintroduction generally that could be agreed to by everyone. Rather, the significance of horse behavior and reintroduction itself was created between particular people in a dynamic manner. To understand what one group of people said about the horses or reintroduction, one needs to go beyond translating the meaning of what is simply said, and to explore the ways in which their meaning is created dynamically amongst their group.

More specifically, social interaction and communication about the horses for each group indexed (Silverstein 1976) different sets of values, i.e., individual or group values. Speaking and writing among scientists are forms of social interaction, and, like other forms of communication, entail their broader values of social interaction. Values are relevant to understanding the meanings that all people involved with reintroduction find in the horses and these values can contribute to the resulting miscommunication that occurs. In this chapter, I explore how international and Chinese project staff brought their different values into the design of monitoring systems and how these values lead to conflict when developing the monitoring system eventually used in Kalamaili.

After 2006, there were several releases of horses, and staff said they would eventually like to establish a total herd of over 400 horses. The horses' reproduction and their increasing population made monitoring a great challenge for Chinese project staff. Internationally, there are different types of monitoring systems for species reintroduction projects. There are scientific monitoring systems that utilize only scientific methods, ranger-based systems, and participatory monitoring systems that formally employ nonscientists. The Chinese project was using a ranger-based system supplemented with the work of Chinese and foreign scientists; the project director hoped that rangers would eventually learn to use scientific methods in their monitoring. As monitoring of these released horses was still below international standards, foreign biologists like Wendelin Schroeder began to advocate for a participatory monitoring system that would formally incorporate local Kazakh herders. As mentioned, I was particularly interested in the distinctive concepts and categories that Kazakh herders might bring to the observation of horse behavior, and I secured the support of the U.S. National Science Foundation for inquiry into this question.

In the vicinity of the Kalamaili reintroduction sites, there are approximately 100 Kazakh households. In a participatory monitoring system, a selection of these Kazakh households would be compensated for patrolling and recording information about the horses, and they would then communicate their findings to rangers and scientists. It was my background as a specialist on Kazakh pastoralism and traditional land use that led to my involvement in the Kalamaili reintroduction. Kurt Austerlitz had originally contacted me for information regarding the Kazakh population, and I later came to be a consultant for their efforts to improve monitoring in Kalamaili. Being so close to the project allowed me special insight into the pluralism—here value pluralism—which is present in international scientific practice. My own participation also forced me to reflect upon my own values that I brought to the project.

Advocacy for any particular monitoring system entails not only understanding the technical requirements and specifications of species monitoring, but also a particular set of cultural values. Kurt and I came from different cultural backgrounds from the Chinese staff, and using our own Western values, we chose monitoring systems that seemed more or less right for us. We thought that participatory monitoring could strengthen monitoring, bring a cutting-edge aspect to research on their behavior, and be more socially inclusive. Chinese staff disagreed with us, and instead, suggested that they would prefer to concentrate on the ranger-based system. They did not offer any technical explanation of why they considered the participatory monitoring system to be ineffective. Of course, each monitoring system has its technical advantages and disadvantages, but as I would learn, the choice between one system and another was closely related to each group's particular values. Within the context of decisions about the design of monitoring systems, each person, including myself, attempted to achieve different value-laden goals through making decisions about that system. The design of a monitoring system was never simply about the horses and monitoring.

This is important when much of the public as well as scientists themselves assume science to be a system of facts and factual statements. In this assumption, the difference between objective facts and subjective values, and the ability to distinguish between them, is thought to be a given. In this assumption, facts state "what is" and what can be discovered by an individual's use of reason and the senses, whereas values say something about what "ought to be"; they are judgments about a future state of affairs which are conventionally agreed to by a group. A simple factual statement might be worded as: "Rangers work in pairs due to policy restrictions and safety precautions." But in contrast, a value statement seems to make a claim about what ought to be—a future preference:

"Przewalski horse monitoring should be done by two observers using scientific monitoring methods." Yet, as I will continue to show, this opposition between the world and our subjective understanding of it and the realms of subjective preference and feeling is highly problematic. The historian of science Lorraine Daston (1995) argues that the uncritical fact-value distinction obscures how the scientific production of so-called facts takes place within a context of action directed towards a value-laden end. Indeed, this is what collapses the fact-value distinction.

The problem for me was not simply one of describing or explaining these plural value systems, but as a consultant for the project, I had to figure out how to deal with problems that emerge between different value systems. In problem-solving value pluralism, I utilize the works of Michel de Montaigne and others influenced by him, such as the philosopher Isaiah Berlin. Montaigne was a sixteenth-century philosopher who thought that resolving conflict between people with different value preferences was less about discovering a singular absolute truth that might support or refute one point of view or another, and more about the intellectual labor of creating a more tolerant life that might include each view. For Montaigne, the goal was not ethical relativism, but finding what works for coordinated action among people with diverse values without enforcing one set of abstract values and suppressing another. Within the context of conflict, I shifted to trying to understand the perspective of my Chinese colleagues, and I found Montaigne's ideas to be useful. For Montaigne, arguments concerning the value of something cannot be stated in ways that are true for all cases, but only to the values they have for particular people in particular cases. In the next section I critically describe my own values as a participant and the move to discuss those of Chinese staff, and most surprising of all, local Kazakh herders.

PARTICIPATORY MONITORING AS SOCIAL JUSTICE

As the anthropologist who now gets directly involved in the story, I would like to offer some background information about myself. In the past, I have researched and written about Kazakh communities and their traditional forms of knowledge as well as investigated the increasing role and importance of modern science in the Kalamaili region. There are approximately 1.2 million Kazakhs in China, and like minorities throughout the world, they face significant obstacles such as discrimination. Because of my experience, I was consulted by the Kalamaili project's foreign participants to assess and advise the reintroduction project. Biologists like Wendelin recognized how the local Kazakh population were vulnerable to the negative consequences of reintroduction and sought to promote a socially sustainable project. After I was asked to provide advice to the project, I thought this might be an opportunity for Kazakhs to be recognized as a valuable group whose extensive local knowledge could make important contributions to Chinese and international science. My original project proposal to the U.S. National Science Foundation was based on this idea of describing and explaining what kind of observations Kazakh rangers would provide to the monitoring project and how these inputs potentially differed from existing observational documentation in the extant scientific literature.

Our choice of participatory monitoring was based on our understanding of the character of local observers, their epistemological and linguistic abilities, and on their inequality vis-à-vis China's ethnic majority and the Chinese state. Participatory monitoring is meant to achieve both scientific and social objectives such as justice, a condition of fair and reasonable relationships (Shiel and Lawrence 2004; Singh and Milner-Gulland 2011; Holland and Blackburn 1998). The method recognizes the social problems inherent in top-down approaches to science that exclude nonscientists or impact them negatively. For example, an Arabian oryx reintroduction in the 1980s did not include local communities nor did it deal with the socioeconomic impact of reintroduction on those communities. As a result poachers decimated the reintroduced population of animals, which were often sold in markets, and ultimately, the reintroduction failed (Spalton et al. 1999). Beyond their instrumental role in conservation, participatory monitoring systems are meant to reaffirm the importance of local communities and to recognize their responsibility in local resource management. They are supposed to take into account peoples' authority over their own natural resources and their right to take part in reintroduction decision-making: these two basic principles are basic to conceptions of justice as well as successful reintroduction policy. Therefore, one of the intentions of participatory monitoring is this explicit goal of social justice. When it works sucessfully and allows people formal roles in reintroduction projects, scholars believe that the livelihood of the community can be strengthened, the quality of management and monitoring can be improved, and community solidarity with conservation goals can be achieved (for a contrasting ethnographic example see Nadasdy 2003).

In my evaluation, I endorsed participatory monitoring. The values I mention—for a socially just reintroduction—informed my preference for participatory monitoring, although I only realize this in hindsight. My values include an ideal society where people come together to decide on issues collectively, and they coincide with the liberal condition described by John Rawls (1971). This condition is an ideal based in notions of individual rights and expresses how things ought to be. In Rawls' view, social justice depends upon a set of voluntarily binding moral obligations, an intimate cooperation or friendship between individuals. It is a voluntary scheme that follows a principle of free and equal cooperation. Rawls says,

> No society can, of course, be a scheme of cooperation which men enter voluntarily in a literal sense; each person finds himself placed at birth in some particular position in some particular society, and the nature of this position materially affects his life prospects. Yet, a society satisfying the principles of justice as fairness comes as close as a society can to being a voluntary scheme, for it meets the principles which free and equal persons would assent to under circumstances that are fair. In this sense, its members are autonomous and the obligations they recognize are self-imposed. (1971, 13)

Within Rawls' idea, there is a sense of the autonomy of each individual and the idea that the obligations of one individual to another are established through agreement and belief. This concept can be applied to anything that has a social dimension—including scientific practice. This value has informed the way I view large-scale projects like species reintroduction and monitoring systems. I value a condition in which scientists and officials invite people into decision-making processes as equals and devalue those conditions in which decision-making is closed to local people, especially when scientists and officials claim a particular kind of authority to make reintroduction decisions that will affect people and their use of land.

The philosopher Hannah Arendt (1993 [1963]) thought that such uses of science had dangerous implications for viable democratic politics, and, by extension, for justice itself. Rather than requiring sharing decision-making with local population, state representatives and scientists believed science to be free from any particular interest and thus underemphasized democratic discussion and dialogue with local populations. This situation in Kalamaili provided an example bearing out Arendt's thought that science has the potential to prevent democratic discussion and dialogue where divergent sides offer diverse perspectives and collective decisions have to be fostered.

For Kazakhs in Kalamaili, land and land access are critically important resources. While they have an ancient and cherished way of life, Kazakhs in China, as a numerically disadvantaged ethnic minority, face a number of problems related to their lower status, second-class citizenship, and lack of power. Many rural Kazakhs have an income level that makes the acquisition of basic food, clothing, shelter, education, health care, and social security central concerns of their everyday life activities. While opportunities for employment and political leadership exist, Kazakhs face significant obstacles such as discrimination in accessing waged labor and maintaining the power to guide their own community affairs. Within this social context, land and access to land are critically important basic subsistence resources. My concern was precisely that the Chinese government did not seek local perspectives on reintroduction, but rather believed itself capable of and obligated to make the appropriate decisions for their interests.

After I became involved, I saw that the situation could not be considered "a scheme of cooperation which men enter voluntarily in a literal sense." From a Kazakh perspective, reintroduction had been imposed on the local population. I discovered that my own support for participatory monitoring reflected my own values, a preference for an imagined state of equal affairs and society. While we do not always acknowledge it, such imagined states of affairs and ideal societies are an important part of the actual practice of science for all participants. In the following account of Wenjun Li, a government official in China, the acceptability of the existing monitoring system and reintroduction was as much related to its effectiveness in monitoring as it was to the real and ideal social interaction it attempted to enable for him too.

RANGER-BASED MONITORING AND A SOCIALIST SOCIETY

Wenjun Li was the director of the Wild Horse Breeding Center (an office of China's regional Wildlife Management Bureau) and head of the Przewalski horse reintroduction, and he had a different set of values from me. He was originally from central China and had relocated to northwest China for his employment. The bureau is located in the regional capital of Urumqi, not far from the city's downtown business district, and is on the eighth floor of a white-tiled building. When I met him, he was wearing a camel-colored sport coat and well-polished shoes, his face tanned and his hair neatly trimmed. On his office wall, there were publicity photographs of Gao Jianxin, an entertainer for the Chinese Community Party and the army. Jianxin was shown clad in military uniform and flanked by two Przewalski's horses. Wenjun thought that monitoring was the responsibility of the state, not the local community. He rejected my recommendations for a participatory monitoring system and was committed to improving the existing ranger-based system by increasing the role of scientists in training rangers to use new methods and equipment. In fact, he saw the rangers' use of local people as a sign of problems within the ranger-based system, giving this as a reason not to focus on participatory monitoring.

People like Wenjun had socialist values that they used to evaluate what would constitute an acceptable monitoring system. He was a member of the Chinese Communist Party, and was familiar with the ideas of leaders like Mao Zedong, Deng Xiaoping, and Jiang Zemin who had extensively theorized the role of both the state and science in an ideal society (Jiang 2000; Mao et al. 2008). The ideologies of these leaders were used to develop a backward society that was under-served by science and technology. Currently party thought understands China as being in an early stage of socialism in which there are still widespread inequities, but which is developing towards more advanced stages of an egalitarian socialist society. Within the party's view, science is essential to achieving this egalitarian goal: it is seen as a productive force in the economy, as a force for management of that economy in relation to natural law, and as the most objective knowledge free from subjective values. After talking with a range of people in China about science, I found this philosophy to be widespread. Moreover, it offered Wenjun a different sense of justice, and importantly, a different sense of how monitoring should be conducted and improved.

In his office, Wenjun gestured to me to sit down, and as we sipped hot green tea out of small paper cups, we talked about the proposed participatory monitoring project. I asked him why he opposed a monitoring system that was participatory, particularly since local people were already participating, albeit informally. Wenjun said,

> Our government staff and herders clearly do not have equal abilities to monitor the horses. Our staff has much more education and knowledge than local herders. They are paid to monitor the horses, and have extensive experience tracking them. Furthermore, the land is technically the government's and herders simply lease it from us. Through development, improved knowledge, and new technologies, herders will slowly increase their education, their knowledge, and their resources. To achieve this, they will be relocated, and be taught new kinds of agricultural practices. Working together, we will achieve a high level of development.

In his view, social and political conditions (as well as scientific abilities and values) were determined by economic conditions. He did not see values such as justice as being particularly efficacious outside of a particular economic system and level of development. This idea has its roots not in liberal thought, but in Marxism; Karl Marx argued that people enter into relationships independent of their will, values, or knowledge. They enter into these relationships through the determination of an economic system. In his view, the processes by which goods and services are produced, distributed, and consumed determine how people interact. These processes include technologies that play a part in determining the degree of justice present in social relationships. Wenjun thought that the continual state-directed introduction of advanced forms of science and technology would eventually, and quite naturally, lead to a more just society. For him, this particular system of values made the ranger-based monitoring system acceptable, preferable, and just.

Wenjun believed that it was the role of the party and state, with their particular resources and expertise, to promote the well-being of the Kazakh and other communities of Chinese citizens—objectives that would be achieved through expert services like species reintroduction and nature conservation. Such initiatives offered a way of making Chinese society healthier, safer, and more sustainable. For Wenjun, this was the only way their leadership would be truly accountable to local people. Wenjun continued,

> We are trying to shift away from traditional ways to a more modern, scientific system. To achieve that, we need to develop our personnel and technology and that requires additional resources. We would like to concentrate all available resources on this change, and ask foreigners to invest their time and resources to help us. In the future we will be a guide and model for conservation in China and all Eurasian reintroductions. . . .
>
> In our current state of development, we need scientific methods and new technologies. Up until recently, we were monitoring the horses under conditions that are much like those of the Kazakh herders. We used tents and had to bring our provisions and supplies into the reserve each time we searched for the horses. We had very high stress levels. Working in the reserve is dangerous; if our truck broke down, we would have a very serious problem. We don't want to go backwards. Things are better now.

For Wenjun, pursuing scientific management with personnel who had received a high level of training and who were capable of using various types of technology (such as satellite monitoring) was equivalent to staffing state hospitals with trained doctors or staffing schools with trained teachers. In fact, the notion of a scientific system for Wenjun was synonymous with a good system, a system in accordance with nature and human needs. He continued,

> China is rapidly developing. We are continuously developing new skills and technologies. Right now, I am communicating with an Italian company that produces weatherproof, motion-activated cameras, and unmanned aerial surveillance equipment. In the future, the countryside will be managed well, and we'll have no problem monitoring the horses. We'll be able to monitor them from here, from our desktop computers in the city, and the work will be comfortable and safe.

He pointed to the city outside his office window, where the downtown area of the regional capital could be seen eight floors below; cars and trucks drove by, mirrored buildings and residential complexes rose high above the ground. His vision was a utopian conception of a futuristic technological society that was fully urbanized and surrounded by protected natural areas teeming with wildlife. In this new society, there were no Kazakh herders, but only urbanized, educated, Kazakh residents. Wenjun's vision transcended what is and all that could be apprehended via direct observation: it was rooted instead in the realm of values, of what ought to be.

VALUE AND ADEQUATE MONETARY COMPENSATION

Kadai, a Kazakh herder who lived around the Kalamaili reintroduction, surprised me by bringing a very different value with which to gauge the participatory monitoring system and the reintroduction of Przewalski's horses more generally. While China was once radically communist, it began a reform of its planned economy in 1979. Since then, China has explored a market economy and has been developing its distinctive form of market socialism. As in other market economies, the development of the market in China led to an increase in inequality, and Kadai, who was quite poor in relation to rising Chinese income levels, had not seen the same benefits of development as other Chinese citizens. In fact, he was struggling financially. Thus, Kadai's logic brought the market value of labor into his evaluation of participatory monitoring and reintroduction.

My original interaction with Kadai began when I saw his name on a list produced during our assessment of those herders who were interested in participating in monitoring. This assessment had aimed to sample the degree of support for participatory monitoring among the Kazakh population. During the assessment, Kadai said that he was interested in participating, and was particularly attracted by the income that he might receive through participation. Later, after the project was rejected by the Chinese government, I sought out Kadai to hear in more depth about his perspectives on the project itself and the government's response. My Kazakh language listening comprehension is good, but I worked with an assistant to clarify the ideas Kadai expressed in the Kazakh language.

When I first met him outside his house, a small three-room mudbrick structure in a village called Karabulgen, Kadai told me he was sixty years old, and had been a herder his whole life. He wore a grey Zhongshanfu, a style of clothing made popular by Mao Zedong, the former party leader. The garment has long been out of style in urban areas, but it is still commonly worn in rural villages like Karabulgen. Karabulgen was a poor village by Kazakh rural standards, made up of approximately one hundred households, mostly mudbrick homes like Kadai's with livestock enclosures and small gardens. It was located just outside the northern border of Kalamaili.

Kadai led a life that was focused on basic subsistence. In Karabulgen, the elderly and school-aged children lived year round in the village, while their relatives moved the family livestock over a series of seasonal pastures to generate foodstuffs as well as products for trade. After leaving Kalamaili, where he had a pasture adjacent to a second reintroduction site, Kadai and his wife would stop in Karabulgen on their way to his spring pastures. In terms of their standard of living, his family met their basic subsistence needs in food, clothing, and housing, and had access to water, land, and livestock. Kadai and most of his neighbors sold whole unbutchered animals to traders and this allowed their families to access enough cash to meet their household needs and to buy basic consumer items like radios and TVs. Nevertheless, their income barely reached the Chinese national average income of $312 per month in 2009, and was only a fraction of the income of urban residents in cities like Urumqi or Beijing. Most of Karabulgen's residents only had eight years of schooling and the most basic levels of literacy.

Monetary value increasingly framed many aspects of Kadai's life. Within the local community, the economic opportunities for him and his family were quite limited, and he had come to accept that he would be dependent on his livestock for the rest of his life. Therefore, while originally expressing interest in the project, he had ultimately come to reject participation in the project as well as reintroduction more generally. He related the work required of him in participatory monitoring with the market value of compensation. Market value, as I use the term here, refers to the average monetary price that labor and other commodities are exchanged for on the market. Kadai had to accommodate the fluctuating market value of the livestock he produced as well as the rising costs of the commodities like clothing and medicine that he consumed. Now, the participatory monitoring project had also put a monetary value on his contribution to Przewalski horse reintroduction. As I asked Kadai about this, he said,

> We were definitely interested in working for the project. As you know, the government has done a lot for us, and so we help the government when we can. But, as the project was designed, it would be very difficult for us to participate. While we are in Kalamaili in the winter, we have many other things to do. To stay there in the summer, it would be very hot and we would need water. We would have to hire people to take care of our livestock or purchase hay. The salary that you said we would get is simply not enough; we wouldn't even break even. We would need more money to participate.

As John Dewey (1981 [1939]) observed, the word "value" can also function as a verb, meaning to appraise, to put a value on, or assign a value to; value can be understood as a rating or a comparison. Anthropologists view market prices as values, judgments about the worth of something, that are specifically embedded in market economies. In other words, values have both qualitative and quantitative characteristics that depend on the social context in which they appear. In Kadai's terms (and context), he compared participating and not participating in terms of monetary costs and benefits; he drew attention to the low monetary value of participating in the monitoring project. As a semi-subsistence herder, monetary worth was an intrinsic part of his value system, rather than broader liberal or socialist ideals.

The historian of science Wolfgang Lefevre (2005) argued that science involves labor understood as production in the sense of a material labor process. Indeed, skilled employment that requires scientific knowledge is an increasingly important component of developed economies. When I was assessing interest in the project, I had originally proposed ¥4000 RMB (approximately $650 USD) per month as an appropriate payment for those monitoring the horses. This was a salary equivalent to that of many urban teachers and medical personnel, and so it was both higher than the average income of Kazakh households like that of Kadai as well as being high in relation to national per capita income averages. The proposed salary for Kazakh herders was designed as a wage, compensation given in exchange for monitoring work conducted as full-time employment, but it did not adequately reflect the reproduction of labor power, acquisition of food, clothing, and shelter within the culturally specific terms of pastoral nomadic Kazakh whose families were dependent on their livestock and pasture. Kadai continued,

> We have helped with finding the horses in the past – in fact, the government said they would compensate us for our labor, but they never did. You have to understand. . . . We use our time to help with the horses, and we need that time to take care of ourselves and our livestock. I cannot just live off these horses like the scientists and government are doing.

The idea of "living off" the horses is an important one as it shows how Kadai was considering the livelihood of his extended family. The reintroduction of Przewalski's horses entails significant sources of capital, in terms of the international translocation of animals and other costs, and has led to the creation of various forms of employment for individuals. This is, in a way, the labor of science. For example, rangers involved in this scientific labor were paid ¥2,500 RMB per month (approximately $400 USD—less than the sum offered to herders to participate), but rangers had all their expenses paid including food, board, and vehicle usage. Rangers often had an apartment in the regional capital that was also provided by the state, and had spouses who also worked for a wage on the job market. The lives of semi-subsistence herders were very different than those of the rangers. They had no wage and all of their income and property depended upon selling of livestock.

Herders would be taking on a very big economic risk by participating in the project. While the wage initially seemed adequate in relation to that offered to rangers, I did not take into account how project conditions would affect the existing Kazakh subsistence strategy of herding and incur new costs. Despite their employment, Kadai's family, not just himself who would be hired, would still require livestock and pasture. In this perspective, the ¥4,000 RMB that was offered for a wage was wholly inadequate to cover the time, food, shelter, transportation, and other costs (including labor and hay for livestock left in the village) that would be required by his extended family. The cost of feed alone for the livestock left in the village would eat up half this wage. Whereas labor compensation was based on a model of individuals, life in Kalamaili was dependent on extended kin groups.

Therefore, Kadai inevitably used market value to judge the project and all its potential monitoring systems critically. He explained,

> There are many problems here. The government takes our pasture to give to the horses. They are probably out there eating our pasture right now. The government has used some of the best pastures, and we are left with the poorest. Their Przewalski horses get fatter, but our Kazakh horses just get thinner and thinner. Yes, they have built up beautiful enclosures at the sites. No doubt—they have used concrete foundations for the fence posts and a nice large door with a lock. Hay is piled up in nice neat stacks. Recently, I heard that they will even dig a well with a petroleum powered pump. If only I could do this for my horses!

Again, he shook his head. Those of us associated with the reintroduction appeared to Kadai to be like wealthy herders making our living, our money, off the Przewalski's horses. His description shows how he reduced the conceptual difference between himself and the scientists—both were herders who simply herded different types of animals. In fact, domestic horse breeding was very much a specialty for particular Kazakh households. So, of course, to herders like Kadai, the reintroduction biologists involved with Przewalski's horses were just a particular kind of herder. He continued,

> Now, they are taking people's land use certificates. They tried to give me a small amount of money for compensation in return for using my pasture for the reintroduction. But I refused. This land is very valuable to us. When we tell them what it is worth, they want to argue. But if we have no pasture, how are we going to be able to live? If my land is taken over by the horses, where will my sheep and cattle go? I am worried about my subsistence. If I have no land for grazing and no livestock, what will I do? They have offered us no alternative. The whole thing is crazy.

Drawing on E.P. Thompson's (1971) classic concept of a moral economy, the political scientist, James C. Scott (1977), has placed subsistence strategies at the center of his understanding of values. For Scott, the threat of basic food shortages explains why rural, agrarian people express certain values. Rural semi-subsistence herders are in a similar position to farmers in that they are exposed to significant livelihood risks, and thus, they will problematize actions that potentially threaten them. Scott's theory helps to explain why Kazakhs like Kadai have the monetary moral values they do. Most Kazakh adults, Kadai included, have experienced food scarcity. So, rather than take undue risks with involvement in reintroduction biology, they make judgments and decisions that seek to avoid such risks; in a sense, their subsistence and lives depend upon being risk-averse. To put it simply, they negatively evaluated the participatory monitoring project, and the reintroduction, entirely because these presented risks that could potentially undermine their security. In so doing, they introduced an unexpected value into the already plural mix.

CONCLUSION

And so, Kazakh herders never did produce descriptions of the social behavior of the horses in their Turkic language. Rather, our attempts to bring Kazakh herders into monitoring ended in conflict between competing sets of values. In this chapter, I have considered how my own experience and my personal ideas of the value of social justice were brought into the decision to plan for a participatory monitoring project. This value now seems abstract and naïve and reflective of my own Western middle class upbringing. I have illustrated how the socialist values of Chinese management staff influenced their negative evaluation of the proposed participatory monitoring system, and their decision to advocate for what they called a "scientific" system. While I see problems with these values, I also empathize with other aspects of them. Finally, I discussed the values of local herders who had at first agreed to discuss the planned monitoring system, but who had their own reasons for later negatively evaluating both the participatory monitoring system and the government system. My experience clearly shows how their decision rested on the values inherent in their own subsistence style and the constraints of their domestic economy. Understanding their values helped me realize that I would rather allow our attempt at participatory monitoring to fail rather than have put them at the risk they perceived. This pluralism in values is an effect of the diversity of international reintroduction biology and of the interactions of three distinct groups of people and is a resource for everyone concerned.

Miscommunication in Przewalski horse reintroduction in China is the result of an inability to develop a solid framework that is capable of equally respecting the different values of foreign scientists, Chinese officials, and local people. Philosopher Isaiah Berlin (1953), drawing upon the ideas of Montaigne, considered several important implications of value pluralism. He contended that no one value system can mediate such a conflict, precisely because the differing values involved are not realizable in a single-system resolution: no one system discussed here can give completely equal respect to such a variety of values, given that each system is specific to only one or a few of the values involved. These are the preconditions that engender what I am referring to as "pluralism." Each system's calibration is predicated on the values it recognizes and engages, so within any one system there is no room to fairly incorporate any of the other systems' values. Some kind of systemic bridging structure has to come into play.

To advocate exclusively for one's own values and to present simplistic choices between and right and wrong, or good and bad is equivalent to eliminating the possibility for value pluralism to exist. As my consultancy continued, and as I continued to converse with Wenjun and Kadai, I slowly began to perceive the substance of other values that were being used by them in their evaluation of the project. I saw, to quote Berlin, "the many sidedness of the truth and the irreducible complexity of life that rules out the possibility of any simple solution, or the idea of a final answer to any concrete problem" (Berlin 1953, 92). In the next chapter I work with the idea that biologists should pay attention to the complex sentiments, concerns, feelings, and emotions that their work may generate or provoke amongst scientists and nonscientists as part of developing more complex—albeit never final and complete—solutions to the problems they face.

FIVE

Sentiments and Assessments of Reintroduction

Even though they may have different value systems and may imagine a better society in different ways, many of the scientists and nonscientists involved with Przewalski horse reintroduction are trying to make society healthier, safer, and more sustainable. Yet, science is not simply a rational control and management of nature and society. While both the theoretical, more abstract knowledge of nature and its practical management play a part in motivating reintroduction biologists, a range of other individual and social needs, desires, and feelings permeate not only the lives of the people who participate in reintroduction but also the lives of those who are affected by it. A broad spectrum of feelings are inseparable from a range of assessments that people make in relation to reintroduction.

The assemblage of things that I have been calling "reintroduction biology" includes human emotions and feelings. Emotions play an important role in how people know and understand the world. For example, scientists describe and experience the horses, or project personnel proscribe or advocate a particular monitoring system, not just from their own reason or sense perception, but from what they feel too. In British philosopher Robert Roberts' (2013) view, moral sentiments figure as "concern based construals," perceptions which are colored by values. For Roberts, concerns are construed in particularly motivated, embodied, and felt ways. Like values themselves, concern based construals are plural, and they reflect a "sentimental pluralism" (Gill 2008).

This chapter explores sentimental pluralism within the context of the Mongolian reintroduction at Hustai Nuruu, a site a thousand or more kilometers northeast of the Kalamaili reserve. As I worked on the participatory monitoring project in Kalamaili, biologists like Wendelin Schroeder and Kurt Austerlitz recommended that I visit the site of a successful participatory monitoring project. They viewed Hustai as a successful project. At Hustai Nuruu I met with park administrators, scientists and rangers, and local people, and discovered that, while successful in establishing a participatory monitoring project (and actively producing scientific knowledge), the reintroduction was fraught with a wide variety of intensely felt emotions. People felt different senses of solidarity and progress, and some had had experiences of betrayal and hardship in relation to reintroduction. They felt the project was good or bad, right or wrong, or some combination of the two. When a range of people offered me their estimations of the nature and quality of the reintroduction of Przewalski's horses, they exhibited what I came to think of as "sentiments" or "concern based construals."

Within the context of science, such sentiments are often seen as simply subjective and thus as an obstacle, rather than a means, to the production of knowledge. Yet, contemporary philosophers of science such as Jeff Kochan (2013) argue that emotions help focus human attention on empirically valid phenomena, to produce lines of inquiry regarding them, and to form conclusions based on reasoning and evidence. In this sense, attending to the role of emotions in science helps me further bridge the dichotomies of subject-object, fact-value, and rationality and irrationality that obscure our understanding of the true complexities of science.

During more than three months spent traveling around the park and surrounding areas, I tried to speak with as many different people as possible who were either involved in or affected by the reintroduction. I worked with an interpreter named Byambaa Dorjatov, a graduate student at the National University of Mongolia, who met me at the capital's main train station, an overnight ride from Beijing. After assisting me in hiring a UAZ469, a type of Russian Jeep, we drove to the park on an excellent paved road. I looked out of the window, listening to Mongolian pop music on the Jeep's radio and watching the countryside pass by. Occasionally, we passed a nomad's truck loaded with a large assortment of objects such as tables, personal belongings, carpets, huge folded pieces of felt, and the wooden structure of the Mongolian nomadic tent or ger. After my arrival and spending time talking with a variety of people, I learned that deeply held concerns and emotions clustered around the use of the Przewalski horse reintroduction for economic development.

As it possesses both natural and cultural resources, Hustai attracts large numbers of international tourists, and it has become an important site for tourists interested in exploring "wild" Mongolia. Hustai Nuruu itself consists of the grassy foothills of the Khenti Mountains and a portion of the rich Tuul River Valley. Like the steppe ecosystems of northern China, the region also has its own endemic species of vegetation, such as the well-adapted Caragana and Amygdalus shrubs (Wit and Bouman 2006). This rich pastoral area has been used by nomadic peoples for millennia, and so the archaeology of Hustai is correspondingly rich too. The impact of the tourist economy was obvious to me as soon as I arrived; there was a paved parking lot, a gate decorated with an artistic drawing of the Przewalski horse and a map of the park, a number of Mongolian gers on concrete foundations, tourist accommodations, and souvenir shops. Behind these, there was a park administration center and housing for scientists. This center also had a few hotel rooms, hot running water, and a resort-like setting with billiards rooms and a restaurant with table linen, silver, and a well-stocked bar. It appeared to be prospering and presented an interesting contrast with Kalamaili.

As I spoke with the park director, scientists, and park employees such as wildlife photographers, I learned about the scientific research and project management for the reintroduction. As he measured the weights of vegetation samples, one employee told me about the project's biological research station, describing its small kitchen, sitting room, computer laboratory, office space, and library. The station was staffed by half a dozen Mongolian staff from various parts of the country. They had studied biology at the graduate level in the urban capital, and some of them had spent time studying abroad in Europe, and had worked there too. Their scientific work was supplemented by the participatory monitoring that was carried out by local nomadic people who had been hired as park rangers. As I traveled throughout the park and the surrounding three counties I learned about how people like these staff, as well as nonscientists and local people, expressed a diversity of sentiments regarding reintroduction. As I listened to them, I could sense how human concerns led to the particular ways in which each of them construed reintroduction.

To begin, I discuss Batyn Tumurlu, the governor of Tsaganuur, and an essential element of the process that made Hustai and reintroduction possible.

EVERYONE BENEFITS FROM THE REINTRODUCTION

The setting for my first example of "reintroduction sentiments" is in the small county of Tsaganuur, just to the south of Hustai Nuruu. The town center there consists of several hundred wooden houses with fenced enclosures as well as a military post, a main government building, and a school. Some of these were built with subsidies during the Soviet period and were quite old. As I wandered through the town, I found older homes with basic dirt floors and minimal material goods, but wealthier residents had built newer homes with concrete floors, kitchens, and Western-style furniture. Within the town itself, there were no concrete sidewalks and the roads were unpaved; I saw people hauling water on wheeled carts from the town well.

Batyn Tumurlu agreed to meet with me to talk about the Przewalski horse reintroduction. Meeting in the three-story government building that towers above the county administrative center, we spoke in her office. A large bright red, blue, and yellow Mongolian national flag hung behind the governor who wore a beautiful Mongolian silver necklace with coral beads. As I got to know her, I found she was an intelligent and articulate speaker, a natural leader. Speaking about the reintroduction, Batyn said,

> Conserving nature to us is very important. Our ancestors worshipped the heavens, the earth, and water. They recognized how dependent we are on nature. But life is different now. Many people are poor and do not have any education. While we need conservation, we also need human development—good nature, good scientific education, and a good economy.

Tsaganuur county gave up almost fifty percent of its land for the establishment of Hustai Nuruu and the reintroduction of Przewalski's horses. This fact was particularly important as most of the county's residents were dependent on county land for their livelihood. The area's participation in Przewalski horse reintroduction, their giving up of their land use rights, was tied to the broad economic objective of reducing poverty. During the Soviet era, local Mongolian governments benefited from large subsidies from the U.S.S.R. which were intended for education, health care, and housing. The government also had guaranteed access to Soviet markets for Mongolian exports, but after 1991 and the collapse of the Soviet Union, the Soviet subsidies were withdrawn and poverty and unemployment rose. The quality and amount of resources available for education, health care, and housing (as well as scientific research) all diminished. Therefore, participating in the reintroduction of Przewalski's horses, and giving up their rights to use land, was intended to stimulate the local economy and reduce poverty through a new kind of international investment. Batyn continued,

> With the participation of biologists and the reintroduction project, we have been successful in developing our community. The reintroduction made investments in building a business center, a new medical clinic, and improvements to the existing school. We showed the community it was in their interest to let go of their land for this broader development. After the community realized the reintroduction was in their own interest, residents left voluntarily. The reintroduction of Przewalski's horses has been very successful here—our economy has greatly benefited.

Mongolia had "transitioned" from a planned economy under on party control to a market economy and multiparty democracy. In this context, Batyn was proud of the project. In fact, she was beaming. We can think of her feelings as concern based construals, in Roberts' sense of perceptions colored by values, or motivated concerns. The central Mongolian government and the local county government of Tsaganuur agreed to cooperate with foreign scientists and they utilized their investment resources to replace past government funding. European and Mongolian scientists met with the Tsaganuur government and convinced them to agree to provide land areas for the establishment of the park. In return, the community would benefit from the establishment of a number of economic development projects. These projects were meant to offset any negative impacts of the loss of land for the county's herders. Scientists promised the local government increased resident incomes and tax revenues as well as direct financial investment in the local economy. Consequentially, Batyn felt positive about the project because it had brought benefits to the ecosystem as well as served the needs of the local population in the new political–economic context.

The development of a national park and a tourist economy was a foreign idea—one that was intended to protect the land used for the reintroduction from other forms of development as well as to grow the local economy. After the Tsaganuur government agreed to the conditions of the project, the scientists, and a number of conservation activists, sought financial support from the United Nations and European governments. Through cooperation, the Tsaganuur government saw international funds channeled into their constrained state budgets, and generally, local residents' incomes improved. In fact, the local economy changed significantly with tourist activities, new handicraft industries, and even small businesses. While the county still had a number of households with low incomes, the average income in the county began to rise. This clearly had an effect on many individuals' and families' sentiments towards reintroduction. It made them, like Batyn herself, feel like reintroduction was a success, that it was a good thing.

In a way, Batyn reversed the idea of who was participating in what; from her perspective, it was scientists and the horses who were participating in Mongolian community development, not Mongolians who were participating in the scientific reintroduction of Przewalski's horses. For her, the reintroduction of the horses was very much about extending material benefits to the population, primarily addressing needs related to income and education, but also to health care and other issues too. Indeed, she emphasized the voluntary way in which residents had given up their land in exchange for other material interests. Again, Batyn commented on this, saying,

> To offset negative effects, the project offered small and large loans to residents and assisted herders who had used the area and had difficulty finding alternative pasture. We offered some of them employment. No one individual was compensated directly, but rather funds were invested in the broader community as a whole, particularly for job creation. Some received jobs as rangers. Others have been hired as cooks, waiters, cleaners, drivers, and storekeepers. In fact, the park is the biggest local employer in the county. Moreover, we organized meetings and allowed the community to participate in the park management decisions. Many participate, and many also get benefits. It really has been a great thing for our county.

Batyn was smiling. County herders were compensated indirectly for sacrificing their land with extensive job creation and thus a better life. Batyn paused, indicating to me to hold on, and answered her desk phone. She then put one caller on hold on her desk phone, and then took another call on her cellphone. When Batyn finished her phone calls, she returned to our conversation. She continued,

> No one in our county opposes the reintroduction because things are much better now. Our schools have received electric pumps for our wells, and so students do not have to carry water to school anymore. Moreover the park has remodeled the student dormitories and provided new beds, blankets, and wall hangings. The park provided sewing machines, DVD players, and educational materials for science classrooms. The park also provided benefits to our hospital; they contributed to improving the heating system and provided use of a vehicle

> as an ambulance. Things in the county are much better now. Everyone supports the reintroduction of Przewalski's horses.

Batyn said she was satisfied with the results of the project. She described Tsaganuur as a unified community with singular interests. In her view, the reintroduction was quite successful in integrating all the different interests of the county within the project's new political-economic forms. This was not only thought by Batyn, reasoned through logic and evidence, but was also felt and affective. If she had different concerns, she might have construed the project differently and land might not have been provided for reintroduction.

Her sentiments reflect a broader ethical consequentialism (Scheffler 1988) that is associated with the project. Consequentialism is an ethical theory that argues that the morality of human action depends only upon its consequences (Duncan and Gray 1979). In consequentialism, as in Batyn's comments, an action is right and good insofar as it promotes happiness among the greatest number of people. As she sees this as her primary goal, the reintroduction can be positively assessed because it is making a majority of the population happy. Biologists often only spoke with people like Batyn, or others who benefited from the project, and formed their own sentiments in relation to them. In fact, there were no formal assessment procedures at Hustai, or within the Tsaganuur government. Thus, these sentiments really are an important part of the practice of reintroduction biology, as important as observation and logic.

REGRET: SOME FAMILIES DON'T HAVE CONNECTIONS

In my second example, I turn to those families who had to give up their land for the creation of the park and the horse reintroduction, but who did not participate in the monitoring system or receive jobs from the subsequent development initiatives. These would appear to be a minority of county residents. Unlike Kalamaili, the Hustai area was used for both summer and winter pastures, and so it was used by pastoralist families twice a year. Approximately 30 households had to give up their land rights in the park and most moved out of the county. After tracking down some of the original residents, I discovered that they had a mix of feelings about horse reintroduction. Unfortunately, I was unable to locate most of the households and neither government nor Hustai itself continued to monitor them after their departure from the county.

From the county capital, I began to travel with my interpreter throughout the outer pasture areas, driving along basic tracks across rolling hills, and stopping at residents' gers to speak with them about their perspectives on Przewalski horse reintroduction. I spoke with a range of people from approximately two dozen households; some were individuals, but most were families. As we drove, a group of ger, or Mongolian tents, would appear on the horizon, and we would make our way towards them. Usually, they were clustered on a pasture, surrounded by exposed earth and livestock dung, and people were engaged in food preparation, livestock management, and other domestic chores. Outside one such ger, I met Demchigiin Khandem. She was well-dressed in a blue *del*, or traditional robe (made from silk and other materials), and she greeted me with a smile.

Demchigiin was part of a herding family who used pastoralism, or mobile herding of livestock, as their essential subsistence and economic strategy. Demchigiin gestured for me to go inside the ger where I saw several people wearing western-style clothes—jeans, turtlenecks, sweaters, and knit hats. They were sitting around a central stove. A man who had been reclining on a bed got up and greeted me. He was named Dash, and was Demchigiin's husband. He offered me a container of tobacco snuff, and after trying a bit, I returned it to him. I told him about my interest in the Przewalski horse reintroduction and the feelings of local residents. He said,

> I remember when the horses first arrived in Mongolia, and people celebrated them. It was wonderful!
>
> Unlike some of our neighbors, I never resisted this project. I thought it might be an opportunity for wealth and prosperity. For progress. But I didn't want to relocate, and I wanted to continue to use the pasture in the park during the spring. Park staff ordered me to leave, but I continued to remain there. It was a very difficult situation with the rangers, and eventually, I left. To try to resolve the problem, I went to meetings at the park office and they eventually persuaded me that leaving the park was a good thing.

Instead of using violence or force, the park staff had used words to convince residents like Dash to support the reintroduction. In their discussions, they had de-emphasized the antagonisms and conflict, the pluralism, that was involved in altering land use arrangements. After Dash was persuaded, his other emotions, and the interests they reflected, changed.

Demchigiin noted that not everyone agreed on whether the reintroduction was good. In her view, the benefits of the project seemed unfair as they were biased towards particular family networks, rather than the community as a whole. In the vacuum left by the collapse of the Soviet-dependent state, kin networks became very important components of subsistence strategies and the broader functioning of Mongolian society (rather than the individuals of market-based economic theory). Kin groups jockeyed for positions in a patron-client like system in which the Hustai park had a dominant role; resources like jobs and loans were given to particular client kin groups. She said,

> Many of the opportunities created by Przewalski horse reintroduction have been monopolized by a minority of the community. Not everyone gets the benefits. Some families have connections with the reintroduction project, and through these, they can secure opportunities. But the project doesn't help those who need it most. There are a number of disgruntled families.

As Demchigiin spoke, she became emotional and I sensed she had a feeling of pleasure mingled with regret, regret for not opposing the project more. Her regret related to how she felt about the ways that things could have gone in more ideal and imaginative directions. She wondered why the people involved with the project acted the way that they did. She considered the feelings of her family. While she and Dash managed to do well through the project, her parent's family and kin did not.

Dash had been listening to his wife, and then he rejoined the conversation. He continued,

> We started working with the director of the park, and he agreed to give us a loan for developing summer tourist activities. We originally came to him with an idea for buying a motorcycle, but he thought that tourist activities would actually bring us money and then we could buy the motorcycle ourselves. Now, he is conducting a tourist market survey, and is planning on ways to increase the number of tourists. We will show them our families' herding activities, things like butter and felt-making. It's a good opportunity. . . .
>
> But not all families have been so lucky. The park director has not been supportive of everyone, including some of my wife's relatives. A number of families have lost their original pasture and have had to make do with nothing.

As Demchigiin and Dash spoke, I realized that the interests, and by extension, the sentiments of Tsaganuur residents are multiple and they reflect a variety of groups of families and kin. "These interests constitute, in the word's most literal significance, something which *inter-est*, {something} which lies between people and therefore can relate and bind {them}" (Arendt 1958, 182). People are simply plural and various; this is why early political pluralists (Cole 1989[1920]; Laski 1919) argued that people are engaged in a vast complex of associations through which distinct groups, with their own concerns and feelings, passions and interests, are formed. Often these emerge from shared occupational work—for example, families that herd livestock together—or use common goods and services—those tourists, for example, that might utilize services at Tsaganuur. Through these processes, they develop shared feelings, and they may come into conflict with other groups around such distinct interests. This is precisely what happened as families who received benefits came to share sentiments and assess the reintroduction in positive terms and those who did not came to assess in negative ones. Demchigiin and her kin were situated between both groups.

In contrast to Batyn's consequentialism, Demchigiin argues that our assessment of reintroduction is not simply about its general consequences for the majority of the population, but its qualitative relationship to all people. It might be assessed in terms of a felt concern for other kin and residents. The philosopher Hannah Arendt argues that actions done out of friendship may have ends, but they are not done for any particular purpose. They are an end in and of themselves. Likewise, Demchigiin sees her relations as an end in and of themselves. So, despite the positive consequences for many people, Demchigiin's overall feelings are much more ambiguous and complicated than those of Batyn, rooted in a very different kind of concern.

THEY ARE MY ENEMY

My third and final example of reintroduction sentiment takes place at a more isolated and much smaller Mongolian ger. As I continued to travel through the county, I often passed groups of ger, usually three or four, with a vehicle, livestock, or some goods around them. But now I found myself at a single ger, with only a patch of bare earth to its side. The first thing that I noticed was a child standing outside the ger: the young boy's hair was knotted and unkempt, and his jean jacket dirty and torn. The ger, just behind him, did not have the single, white canvas covering that I often saw, but was instead covered by several multicolored pieces of old canvas and plastic which had been roughly tied together. This covering barely reached the ground, and earth was piled up around the outside edge. Clearly, it was a very poor household, without associations with the project, and without the benefits and advantages provided by the project.

As we approached the front door and I announced myself via my interpreter, no adult came out to greet us. As I entered, I saw a woman, Munkjargal, who was sitting washing dishes in a red plastic basin. The inside of the ger had little furniture or property, just some horse tack, a small stove, and cooking goods; flies were buzzing around. Munkjargal appeared to be in her fifties, and had reddened skin and greyish hair pulled back behind her head. She wore an old grey, plaid flannel shirt. She did not stand to greet us. After introductions, we asked her about the Przewalski horse reintroduction, and as she spoke, tears began to run down her cheeks. She said,

> Most people only want to say good things about the reintroduction. They think since it is science that is must be good. They are afraid if they say something, and the scientists find out, they will get in trouble. But I am not afraid. Ever since they brought the horses here, my life has

> gotten worse. I lost my pasture, and even though they said we would be compensated, we received nothing. I have had to sell most of my livestock. For me, Przewalski's horses are my enemy.

Munkjargal was angry. She was angry at what had befallen her, particularly since her misfortune was directly linked to the good fortune that others had experienced. She felt insulted and hurt. She felt she was a good person, was important and significant, and that she deserved more from her neighbors and community leaders. Whereas they had acquired good things, she had lost land and endured hardship. In the face of the events which befell her, she felt powerless. Thus, she felt as though the horses were her enemy. Her story provided another example of the way that human life and activities including science has this combination of perception, thought, and emotion.

When they received nothing in compensation, herders like Munkjargal put their feelings into words and they assessed the project in negative terms. They saw the project as a threat to their interests. She continued,

> Whereas I used to be close to a water source, now I have to travel ten miles just to get water. Moreover, the rangers are so antagonistic and aggressive towards us. They constantly chase our livestock away from the park boundaries and the Przewalski's horses. Like this, its hard to fatten what little livestock I have left; the rangers chase them and they run and run and get skinnier and skinnier. At the same time, some of the rangers have huge herds; I heard that Urtnasangin [a ranger] has almost one thousand sheep. He gets to have the run of the land now. When I raise my concerns, they say they don't care. They say the horses are what are important. If you don't believe me, ask Halzan. He is my witness.

Munkjargal thought the project had divided the community into competing factions, and she no longer saw the rangers who were hired for participatory monitoring (such as Urtnasangin) as her neighbors or friends. For her, their passions, interests, and identities were not the same as her own. She continued,

> I was born there and I thought I would die there. My winter pasture was in the main valley of the park, and that pasture was my father's, and his father's too. I loved that pasture, and it has many important memories for me. We were the original people here, and we should decide how the land is used. Not the scientists.

While we spoke, most of the rest of her family remained quiet. As much as her own passions and interests, the land itself was an important component of her individual perceptions. Her memories of the land were an important constituent of her personal identity. Her interests, and her identity as an individual, could not be subsumed by the interests of other associational groups, the broader community, or the state itself: in fact, I began to understand how Munkjargal reflects the human condition of plurality.

I recognize I share an assumption with Western biologists like Wendelin. While pluralism is made up of sets of sets, it does not preclude individuals, even if I might challenge their foundational quality. But, for Munkjargal, there is no way to completely leave aside her distinctive passions and interests. They can never be completely subsumed by other associational groups, either of her kin group or of other displaced groups or individuals. To be human is neither about society's mutuality and common interests, nor the more limited interests of a group. For Munkjargal, her passion is her unique life, body, and voice. It would be unlikely, says Arendt, for someone like Munkjargal to criticize the project, "if {humans} were endlessly reproducible repetitions of the same model, whose nature of essence was the same for all and predictable as the nature of essence of any other thing. {But} plurality is the condition of human action because we are all human in such a way that nobody is ever the same as anyone else who ever lived, lives, or will live" (1958: 8).

This particularity of concern and feeling is essential to understanding the emotions that are prevalent in Munkjargal's assessment of the Mongolian reintroduction.

CONCLUSION

In philosophy, there is a long tradition of viewing the sentiments as that which motivates humans and drives human action. In this way, sentiments were conceptualized like hunger and desire that drove people to act and to interact with others, either to cooperate or express antagonism. A pluralist view recognizes that in all aspects of human life including the practice of science there are multiple sentiments that drive people to different understandings and different forms of action. While positive assessments of the Przewalski's horses in Mongolia's Hustai Nuruu are connected to the positive feelings of many residents, there are a range of sentiments that include feelings of uncertainty and ambiguity, regret and anger. These feelings are just as important as more positive ones as they allow for us to recognize a range of potential assessments rather than simply a univocal positive or negative one. Recognizing sentimental pluralism, we might assess positively some aspects of the reintroduction, while negatively assessing others. It might lead to mixed moral feelings. And these mixed moral feelings are precisely what I explore in the next and final chapter.

Conclusion

Pluralism and Implications for Future Reintroductions

In the twenty-first century, modern societies like China and Mongolia face an array of challenging environmental and social problems. Przewalski's horses are only one possible example of species endangerment, the scientific and popular response to it and its social complexity. Through various projects designed to ameliorate the negative impact of both endangerment and reintroduction activities on society, scientists like Wendelin Schroeder develop relationships with horses such as Bogdan and Chulun, with regional scientists like Dr. Zhen Yang, and nonscientists such as Wenjun Li and Batyn Tumurlu. As a result, reintroduction biology is a complex assemblage of the human and nonhuman, a vast, global enterprise, a set of powerful and life changing forces, that has many implications for understanding a changing nature and society.

At present, there are more than 2,000 Przewalski's horses known to be alive throughout the world—more than have existed during the last 100 years. Seen from this numerical perspective, the reintroduction of Przewalski's horses might be superficially read as a successful example of how scientists and nonscientists have developed their knowledge and applied that knowledge to the resolution of a socially and environmentally based problem. But, as I have shown, a realistic assessment of reintroduction, and the scientific practice of biology behind it, requires a far more complex reading and mixed set of moral feelings. The preceding chapters have demonstrated how anthropological inquiry and philosophical reflection can contribute to a more nuanced understanding of reintroduction biology. To this end, I have suggested using the concept of pluralism to break through unitary and simplistic understandings of its knowledge and practices.

Given that this science plays a key role in interventions to mitigate and stop species endangerment and extinction, both reintroduction projects themselves and the concomitant perceptions of their success will probably become more widespread throughout the world over the next few decades. Reintroduction biology is rapidly developing as scientists estimate that there are over 2,500 other critically endangered species in the world, many of whom could be the object of reintroduction in the future. Similarly, nonscientists are becoming increasingly aware of the impending loss of large animal species like horses, even if many people do not always understand exactly what the scientific practice of their conservation, breeding, and reintroduction entails. An exploration of the broader implications of pluralism in Przewalski horse reintroduction can enable a deeper scientific as well as public understanding of an expanding and increasingly prevalent international scientific practice.

A naïve public embrace or rejection of reintroduction biology and scientific practice is inadequate for the environmental and social problems facing our world today. At the same time, my analysis of Przewalski horse reintroduction questions the assumption that science is inevitably progressing towards a totalizing comprehension and that the activities of scientists will somehow emancipate people from those problems. Based on the evidence of my observations and my previous discussion of reintroduction, I contend that scientific knowledge can neither be said to be simply progressing towards an all-encompassing comprehension, nor is it capable of straightforwardly emancipating people from their problems. Clearly, there are still many uncertainties concerning the horses and, furthermore, the people who work with, for, and around them still face many kinds of natural and social problems. This does not mean that we do not need reintroduction science, but that we do need to learn to think about it in a more nuanced, sophisticated, and creative way. This is one of the reasons why I have utilized the concept of pluralism and have viewed science as an assemblage of different phenomena. As such a complex assemblage, reintroduction has aspects that I advocate for as well as those I am critical of.

As I conclude this book, I tie together and reflect on the big picture of reintroduction biology and its implications as we consider the future of other reintroductions (and other forms of international scientific practice). It's a picture I have sketched and which I qualify as such to be singular and united. This picture includes several basic principles:

NATURAL COMPLEXITY WILL ALWAYS TRANSCEND THE BOUNDS OF A PARTICULAR SCIENTIFIC KNOWLEDGE

As we discussed, there will always be more to animal behavior, let alone larger phenomena such as nature, than any one individual or group can observe and reason about. The philosopher Jason Turner (2010) describes ontological pluralism as the recognition that everything that "is" has different ways or modes of "being." When an observer describes and tries to explain the existence of a thing, they usually recognize only one way or mode of its being. Turner, in thinking as an ontological pluralist, points out that "thinking of reality as having a single ontological structure—a single [system]—is a mistake, as is thinking of ontological categories as divisions within this single structure. Rather, reality has multiple ontological structures—multiple, independent [systems]" (2). Many of these structures are likely to be unknown and unknowable, by one or another individual human or human population; thus, ontological pluralism presents basic limitations on science or any human knowledge and its basis is physical and material, in what actually exists in the world. Scientists need to learn to accept this uncertainty and remain open to the potentially unsettling outcomes of the observation and reasoning of other individuals and groups. In fact, I have shown how the different observations and reasoning of a variety of individuals and groups can be resources for understanding ontological complexity and improving science.

Consider how the observations of one group of scientists—climatologists, for example, affects that of another, reintroduction biologists—even within one dependent system. Atmospheric changes due to economic development and the increasing human usage of fossil fuels (as well as a nonhuman climatic cycle) are trapping solar radiation and leading to global temperature increases. The Earth's existing climates, and the ecosystems which depend upon them are changing too with glacier loss, rising sea levels, and other phenomena. Data from sources such as the U.S. National Atmospheric and Oceanic Administration demonstrates significant warming on land and sea, and the same institution has documented land changes, fresh water scarcity, and species migrations and loss. In such a context, the very reintroduction sites that I have discussed may prove to be unsuitable habitats for the reintroduced horses in as short a time span as fifty years. This changing nature of the sites themselves means that the long-term viability of conducting species reintroduction becomes highly questionable. Turner's idea of reality as constituted by multiple independent systems helps us understand why scientists struggle to create a coherent and comprehensive account of the world, some of them working to reintroduce the horses to their former habitats while others concern themselves with habitat change and the unfeasibility of reintroduction. No scientist nor system of knowledge can integrate all relevant aspects of one system, let alone multiple, ontological structures and systems. This is an inevitable part of the complex nature and structure of our world.

In my discussion of the scientists involved with describing Przewalski horse social behavior, I demonstrated how their knowledge of horses changed as the material, environmental conditions of the horses changed from captivity to the wild. Whereas scientists assumed the horses were unitary objects with particular fixed properties and qualities, they eventually discovered the horses were not a single entity, but a multiplicity of beings that change through time. From objects with fixed capacities, they discovered those capacities emerge out of the interaction of a multiplicity of phenomena. The differences and dissonances between the observations of individual scientists eventually became an important resource rather than an obstacle to their understanding of the material variation in the horses themselves. Moreover, I have discussed how different groups of scientists might even provide mutually distinct, but complementary descriptions and explanations of horse behavior. Scientists themselves increasingly recognized that nature has a far more complex, changing, and diverse organization and structure, and that singular descriptions will always be incapable of producing general and universal explanations. This is a humbling, but realistic, realization of the limits of human knowledge for future reintroductions as well as any form of scientific practice.

AN ADEQUATE BUT IMPERFECT KNOWLEDGE OF NATURE WILL REQUIRE MANY DIFFERENT WAYS OF KNOWING

Reintroduction biology is a vast network of nonhuman animals and human individuals and groups. There are zoo curators, biologists, technicians, government officials, park administrators, and local people dedicated to, or complicit in, advancing the practice. Each is involved with the production of knowledge and practice of reintroduction, and each has important roles in contributing to our knowledge of a complex world. No one individual or group has unqualified knowledge of the whole and knowledge is distributed unevenly across individuals in the network. Only through maintaining this human diversity and its epistemological pluralism, a very large array of methods and approaches that utilize many sources and conditions of knowledge, will natural complexity be better understood. Despite a popular rhetoric of best practices, future work should maintain a commitment to a diversity of epistemological practices.

As I finished my work in Huotai, I spoke with Urtnasangin Tumurkhuyag, a former herder now participating in monitoring. Having basic primary schooling, Urtnasangin worked for a period before Mongolian independence as a mechanic in addition to a broader life of sheep herding. It is precisely this background that makes Urtnasangin an extraordinary scientist, in Kuhn's sense (1996), and allows for him to note anomalous phenomena that fall outside existing biological behavioral paradigms. I asked him about the behavior of the horses, and he talked about how the stallions can predict the weather. The horses predict the weather and then decide where to graze or drink accordingly, he said. For example, if the stallion begins to ascend a mountain and lead the group upwards in altitude in the evening, the weather will be hot the following day. If he guides them into sheltered valleys, the weather will be cold and harsh. He said that his insight did not result from any particular method of observation, but that he had made many, many observations through time. As the philosopher of science Paul Feyerbend has said, "There is only one [method] that can be defended under all circumstances and in all stages of human development. It is the principle anything goes" (2010, 12). In thinking about Urtnasangin's observation, I would agree. Feyerbend calls this an anarchist theory of knowledge and argues that rigid application of a unitary set of methodological rules is liable to inhibit the progress of knowledge production. If scientists were to dismiss Urtnasangin's observations based on his personal background, his education, or his methodology, they would potentially lose access to his unique, interesting, and provocative observation. My recommendation is for scientists to remain open to the already diverse epistemologies utilized in their practice. As Feyerbend notes, people do not gain insight into the world because they possess a magic wand—a golden methodological bullet—but because they have studied it for a long time and gained experience-based knowledge.

Sandra Harding (1998, 62), another philosopher or science, argues that it is precisely because nature is not uniformly organized, that particular individuals and groups do not have access to observation of the entire diversity of nature's features. For example, there were many aspects of the horses that Urtnasangin did not observe and will never observe. This is true for scientists too. People in different locations, both literally and metaphorically, in a diverse nature become exposed to different aspects of it, and those differences in exposure can be a valuable resource for advancing knowledge. For Harding, different cultural groups, seen as different human adaptations to nature, are particularly important repositories for historically developed beliefs that can guide the production of knowledge about nature. Cultural diversity among observers might insure checks and balances across paradigms. The value of the different ways of knowing of differently situated people—particularly culturally distinctive local people living around reintroduction sites such as Mongolian or Kazakh herders—should not be underestimated in terms of their potential to contribute to scientific knowledge. They are more than just the laboring "eyes" and "ears" of distant scientists, as those individuals and groups sometimes think of them, and can contribute to how we scientifically explain nature too.

DIFFERENT HUMAN LANGUAGES ALLOW FOR DIFFERENT CONCEPTUAL INSIGHTS INTO NATURE

Language is a crucial component of culture, although we usually use language with little self reflexivity. There are a number of languages used in reintroduction and this linguistic pluralism relates to the variety of forms of perceiving and reasoning about the horses and their behavior. Written descriptions associated with particular languages can emphasize one or another aspect of horse behavior; German biologists have focused on individual horses, agonistic behavior and dominance hierarchies, while Chinese scientists have focused on horse groups, nonagonistic and cooperative behaviors. As I learned, the languages of scientists are tools or technologies which, in drawing on structurally embedded cultural paradigms, cut apart the ontologically plural world of the horses. They achieve this by guiding an individual or group's observation through a set of linguistic concepts and categories. Pluralism, in this sense, does not simply refer to a material, physical world or how we might come to know about that world, but also to the ways that we communicate between ourselves, as humans, about that world, how we render that world meaningful. Attempts to undermine such diversity through the advocacy of a common language such as English among international networks of scientists is thus a challenge to our knowledge itself.

Martin Heidegger (1971) thought that we might assess our knowledge not just by using one or more languages, but also by reflection on how we use them. My analysis would support this supposition. One of the interesting ways that this recently surfaced was in relationship to research on what Wendelin call "fox coloring," an English vocabulary item, in the horses. All the Przewalski's horses that are alive today (at both Kalamaili and Hustai as well as those in captivity) are descendants of 12 founder animals that were mostly captured in the wild in Asia and transferred to captivity in Europe. She believes that some of the founder individuals might have been interbred with domestic horses as some individual horses born in captivity and the wild have what she said was an unusual "fox coloring." Since the Przewalski's horse had become extinct in the wild, there is no possibility to consult a wild population to understand the natural color variability in the species. Instead, she decided to consult translators and the language of the earliest written species descriptions in Russian language.

In this example of the importance of language and multilingualism in reintroduction, we can also see how it challenges ideas of language as an objective instrument of description. In terms of the language of color, Wendelin originally took the term "fox coloring" as nature itself. A color like "fox" would not seem to be an abstraction, a theory, and a philosophy. Yet, in terms of nature itself, color is very diverse and spectral. As she reflected on the nineteenth century descriptions of the color of Przewalski's horses, she discovered the basic nonequivalence of "fox" color in the Russian language. Rather, Russian-speaking and writing observers used vocabulary items that could not be used in a way that had the equivalent meaning of "a color like that of a fox." As she explored the range of descriptions of the color of the horses, she discovered the degree to which color forms a continuum: red is not different from brown at some definite point and reddish-brown can involve different amounts of yellow, white, and black; moreover, and more importantly, how color appears to an observer depends on a range of reflective conditions of light and is, like all colors, unstable. Across languages, the same word can be used to denote colors that might have two or more terms, and depending on the conditions of light, a description of the color of an object can diverge in a different language. At the time of writing, Wendelin discussed the uncertainty around the color of the horses in the wild, as much a linguistic insight as a physical, material one. Therefore, an exploration of language highlights that there is no perfect language of concepts and categories that maps onto the real nature; just as nature is ontologically plural, scientists need a diversity of concepts and categories to make sense of its pluralism.

INTERNATIONAL SCIENTIFIC COMMUNICATION BRINGS NORMATIVE SOCIAL ORDERS INTO CONFLICT AND REQUIRES MEDIATION

One of the reasons for conflict in reintroduction biology is that differences and divergences, linguistic or otherwise, can lead to disagreements. Scientists, like all people, cooperate with others within a particular society, and are intertwined, at the level of their basic human needs, with them. The theorist Jurgen Habermas (1971) has argued that language, whether in science or not, is part of this social coordination. Language helps a group of people to develop a common understanding and thus coordinate their action socially. Language, in his view, is never only about reference to nonhuman ontological reality, but always has this component of social coordination. Likewise, I described how scientific communication can express what a group of people hold to be important for them as a group, and their values concerning common life and social action, even when the topic appears to be more technical and instrumental. The problem, in the context of international reintroduction biology, is that communication across social groups and societies—often with very different political relationships—and often brings differing values into conflict.

Within the context of the Kalamaili reintroduction, Western biologists and advocates like myself interact with Chinese biologists and officials. China, unlike Germany or the United States, is a socialist republic ruled by the Chinese Communist Party under the principles of democratic centralism. The decisions that governmental and party personnel make concerning monitoring are inseparable from the political objectives associated with its broader scientific and technological policies and under the principles of democratic centralism, must be authoritatively supported and implemented. Thus, statements about monitoring systems, about the need for state-run, expert systems, are also attempts to coordinate a particular kind of social action. As Western biologists and advocates carry different values, their decisions regarding monitoring come into conflict with their Chinese counterparts and lead to miscommunication.

From my experience, understanding the value-laden aspects of scientific practice will be very important for future international reintroductions. As the Western biologists and myself abandoned our attempt at developing a participatory monitoring project, later discussions with Chinese staff highlighted a number of issues. Most importantly, a broad translation of not just vocabulary items, but the values of each group more broadly needs to be an important focus of collaboration, not just the practical matters of scientific practice itself. This clearly will require mediators who themselves are not a part of the existing or potential conflict. Moreover, western biologists must realize the way that historical conflicts between societies—particularly between different ideologies—can also pay out with scientific practice. This is a real challenge to the instrumental and technical focus of scientific practice, and entails changes to the broader public discourse about what science is, the training of scientists, and the actual practice of scientific collaboration. It will also entail deference to the valves of the home country and what pragmatically works rather than an unwavering commitment to one's own values, something that took me a long time to learn.

POPULAR SENTIMENTS ARE IMPORTANT CONCERN BASED CONSTRUALS AND THEIR COMPLEXITY NEEDS TO REMAIN WITHIN PROJECT ASSESSMENTS

Conflicts based in different values can have great emotional energy. Yet, to advocate exclusively for one's own values and to present simplistic choices between and right and wrong, or good and bad, is equivalent to eliminating the existence of value pluralism, and human pluralism, generally. Throughout this project, I slowly began to perceive the substance of the diverse values that are being used by all of the different people involved in the network. I saw "the many sidedness of the truth and the irreducible complexity of life that rules out the possibility of any simple solution, or the idea of a final answer to any concrete problem" (Berlin 1953, 92). This is as true for the color or behavior of the horses, as it is for how groups of people should monitor them, and finally, for how everyday people assess reintroduction too.

No matter what my own values are, I think reintroduction biology is generally a good thing. Reintroduction biology is for the earth's biological diversity and the future of our species, for human knowledge of the biological world, and for various forms of national, community, and individual human development. So, whether or not one values participatory monitoring and the inclusion of local people, the sentiments of nonscientists should be taken seriously in planning and modifying reintroduction. It is, in a very real sense, for everyone, for nonscientists, and their feelings about it matter. Sometimes the horses themselves and scientific practice can overshadow this. Yet, future reintroductions need to attend better to the range of voices of people impacted by its activities.

In both Kalamaili and Hustai, communities have a variety of sentiments regarding reintroduction, and they think about and question the decision-making behind it. Popular sentiments, sometimes deemed irrational in relation to the rationality of science by scientists, are very much a part of the practice of reintroduction biology; they are not diametrically opposed to its empirical observation and logical thought processes, and to the contrary, are an inseparable part of it. The sentiments of nonscientists, as I have shown, relate to their particularly diverse kinds of concern-based evaluations of the impact of reintroduction on their lives. As I began to take nonscientists' diverse sentiments seriously, I became critical of simplistic assessments of the reintroduction project in terms of its "success" or "failure." All too often, scientists' assessments of reintroduction are either univocally good or bad, right or wrong. I began to think that assessment of reintroduction needs to reflect a more mixed set of evaluations and moral feelings. I think recognizing pluralism, at the very least, entails recognizing the ambiguity and uncertainty in the outcomes of reintroduction.

We do not need to hold to the idea of a single reality, a single method to know it, and absolute certain knowledge. We do not need simply shared values and consensus. We can open ourselves to the mystery of a complex universe, multiple, contradictory truths, and many, many ways forward. In conclusion, without unity or certainty, efforts towards dealing with species loss will not fall apart; with pluralism as a guide, scientists and the public might be opened to seeing new things about the species and the natural world, to developing new kinds of understandings, and interacting in new, more inclusive ways.

References

Arendt, Hannah. 1958. *The Human Condition*. Chicago: University of Chicago Press.

———. 1993[1963]. "The Conquest of Space and the Stature of Man." In *Between Past and Future*. New York: Penguin Books.

Berlin, Isaiah. 1953. *The Hedgehog and the Fox: An Essay on Tolstoy's View of History*. New York: Simon and Schuster.

Boellstorff, Tom. 2008. *Coming of Age in Second Life*. Princeton: Princeton University Press.

Bogost, Ian. 2015. "The Cathedral of Computation." *Atlantic*. January 15.

Bourdieu, Pierre. 1977. *Outline of a Theory of Practice*. Cambridge: Cambridge University Press.

———. 1997. *Pascalian Meditations*. Stanford: Stanford University Press.

Cao Qing, Defu Hu, Melissa Songer, and Peter Leimgruber. 2009. "Habitat Selection in Przewalski's Horse in the Kalameili Nature Reserve, Xinjiang China." Restoring Przewalski's Horses: Lessons Learned from In-Situ and Ex-Situ Conservation, International Congress for Conservation Biology, Beijing, China, July 11.

Chen Jinliang. 2008. Fangui pushi yemade shi, shuiyuan, kongjian liyongji shengcun duicede yanjiu. Utilization of Food, Water, and Space by Released Przewalski Horse (*Equus Przewalski*) with Reference to Survival Strategies Analysis, in Chinese. Ph.D. Dissertation, Beijing Forestry University. Beijing: Beijing Forestry University.

Chen Jinliang, Defu Hu, Qi Lu, and Yuping Meng. 2006. "Pushi yema yu jiamade yitong." Differences Between Domesticated and Przewalski's Horses, in Chinese. *Da Ziran* 6: 39–40.

Chen Jinliang, Qiang Weng, Jie Chao, Defu Hu, and Kazuyoshi Taya. 2008. "Reproduction and Development of the Released Przewalski's Horses (*Equus przewalskii*) in Xinjiang, China." *Journal of Equine Science* 19(1): 1–7.

Choy, T., L. Faier, M. Hathaway, M. Inoue, S. Satsuka, and A. Tsing. 2009. "A New Form of Collaboration in Cultural Anthropology: Matsutake Worlds." *American Ethnologist* 36(2): 380–403.

Coady, C.A.J. 1992. *Testimony: A Philosophical Study*. Oxford: Clarendon Press.

Cole, G.D.H.. 1989[1920]. "The Social Theory." In *The Pluralist Theory of the State*. Paul Q. Hirst, (ed). London: Routledge.

Daston, Lorraine. 1995. "The Moral Economy of Science." *Osiris* 10: 2–24.

DeLanda, Manuel. 2002. *Intensive Science and Virtual Philosophy*. London: Continuum.

Deleuze, Gilles. 1991[1953]. *Empiricism and Subjectivity*. Constantin Boundas (trans.). New York: Columbia University Press.

———. 1993[1988]. *The Fold: Leibniz and the Baroque*. Tom Conley (trans.). London: Athlone Press.

———. 1994[1968]. *Difference & Repetition*. Paul Patton (trans.). New York: Columbia University Press.

Descola, Philippe. 2013. *Beyond Nature and Culture*. Chicago: University of Chicago Press.

De Vries, Han. 1995. "An Improved Test of Linearity in Dominance Hierarchies Containing Unknown or Tied Relations." *Animal Behaviour* 50: 1375–89.

De Vries, Han, and Michael C. Appleby. 2000. "Finding an Appropriate Order for a Hierarchy: A Comparison of the I&SI and BBS Methods." *Animal Behaviour* 59: 239–45.

Dewey, John. 1988[1939]. "Theory of Valuation." In *The Later Works, 1925–1953*. Jo Ann Boydston (ed). Carbondale: Southern Illinois University Press.

Dreyfus, Hubert. 1987. "From Socrates to Expert Systems." In *Interpretive Social Science: A Second Look*. Paul Rabinow and William M. Sullivan (ed.). Berkeley: University of California Press.

Duhem, Pierre. 1991[1914]. *The Aim and Structure of Physical Theory*. Princeton: Princeton University Press.

Duncan, Greene, and John Gray. 1979. "The Left Against Mill." In *New Essays on John Stuart Mill and Utiliatarianism*. Guelph: Canadian Association for Publication in Philosophy.

Engels, Friedrich. 1940[1883]. *Dialectics of Nature*. New York: International Publishers.

Feh, C. 1988. "Social behavior and relationships of Przewalski's horses in Dutch semi-reserves." *Applied Animal Behavior Science* 21: 71–87.

Feyerbend, Paul. 2010[1975]. *Against Method*. London: Verso Books.

Foley, Richard. 2001. *Intellectual Trust in Oneself and Others*. New York: Columbia University Press.

Franklin, Sarah. 1995. "Science as Culture, Cultures of Science." *Annual Review of Anthropology* 24: 163–84.

Geertz, Clifford. 1988. *Works and Lives: The Anthropologist as Author*. Stanford: Stanford University Press.

Gerken, Mikel. 2015. "The Epistemic Norms of Intra-Scientific Testimony." *Philosophy of Social Sciences* 45(6): 568–95.

Gill, Michael B. 2011. "Humean Moral Pluralism." *History of Philosophy Quarterly* 28(1): 45–64.

———. 2008. "Sentimental Pluralism: Moral Psychology and Philosophical Ethics." *Philosophical Issues* 18(1): 143–63.

Goodwin, Charles. 1995. "Seeing In Depth." *Social Studies of Science* 25(2): 237–74.

Habermas, Jurgen. 1971[1968]. *Knowledge and Human Interests*. Boston: Beacon Press.

Halliday, M.A.K., and J.R. Martin. 1993. *Writing Science: Literacy and Discursive Power*. Pittsburgh: University of Pittsburgh Press.

Hanson, Norwood Russell. 1958. *Patterns of Discovery: An Inquiry into the Conceptual Foundations of Science*. Cambridge: Cambridge University Press.

Haraway, Donna. 1989. *Primate Visions: Gender, Race, and Nature in the World of Modern Science*. New York: Routledge.

Harding, Sandra G. 1998. *Is Science Multicultural?: Postcolonialisms, Feminisms, and Epistemologies*. Bloomington: Indiana University Press.

Hathaway, Michael. 2009. "Postcolonial Science Studies and the Making of Matsutake Science in China." *American Ethnologist* 36(2): 393–97.

———. 2014. "Transnational Matsutake Governance: Endangered Species, Contamination, and the Reemergence of Global Commodity Chains." In *Mapping Shangri-la: Nature, Personhood, and Policy in the Sino-Tibetan Borderlands*. Emily T. Yeh and Christopher R. Coggins, editors. Seattle: University of Washington Press.

Heidegger, Martin. 1962[1927]. *Being and Time*. New York: Harper.

———. 1969. *Identity and Difference*. New York: Harper & Row.

———. 1971[1959]. *On the Way to Language*. New York: Harper & Row.

Helmreich, Stefan. 2009. *Alien Ocean: Anthropological Voyages in Microbial Seas*. Berkeley: University of California Press.

Herder, Johann Gottfried. 2002[1772]. "Treatise on the Origin of Language." In *Herder's Philosophical Writings*. Michael N. Forster, editor and translator. Cambridge: Cambridge University Press.

Holbraad, Martin. 2012. *Truth in Motion: The Recursive Anthropology of Cuban Divination*. Chicago: University of Chicago Press.

Holland, Jeremy, and James Blackburn. 1998. *Whose Voice? Participatory Research and Policy Change*. London: Intermediate Technology Publications.

Holton, Gerard. 2000. *Einstein, History, and Other Passions: The Rebellion Against Science at the End of the Twentieth Century*. Cambridge: Harvard University Press.

Houpt, Katherine A., and Lee Boyd. 1994. "Social Behavior." In *Przewalski's Horse: The History and Biology of an Endangered Species*. Albany: State University of New York Press.

Huang Xiangyun, Shengnan Ji, Yongjun Zhang, Jie Cao, Hongjun Chu, Kai Li, and Defu Hu. 2012. "Fanggui chuqi pushiyema jiazuqunde biange jiqi guanghu yiyi." Change of Released *Equus Przewalskii* Family and the Implication to their Management in Initial Period, in Chinese. *Sichuan Journal of Zoology* 31(3): 488–96.

Humboldt, Wilhelm von. 1999[1836]. *On Language: On the Diversity of Human Language Construction and its Influence on the Mental Development of the Human Species*. Michael Losonsky, editor. Peter Heath, translator. Cambridge: Cambridge University Press.

Hume, David. 1963[1748]. *An Enquiry Concerning Human Understanding and Other Essays*. New York: Washington Square Press.

Jiang, Nan. 2005. *Yemade Diyu. The Whisper of the Wild Horse, in Chinese*. Kunming: Yunnan People's Publisher.

Jiang, Zemin. 2000. *Lun Kexue Jishu*. Theorizing Science and Technology, in Chinese. Beijing: Zhongyang Wen Chubanshe.

Kant, Immanuel. 2007[1781]. *Critique of Pure Reason*. Marcus Weigelt, translator. New York: Penguin Books.

Keiper, R.R. 1988. "Social interactions of the Przewalski horse (*Equus przewalskii* Poliakov 1881) herd at the Munich Zoo." *Applied Animal Behavior Science* 21: 89–97.

Keiper, R.R., and H. Receveur. 1992. "Social interactions of free-ranging Przewalski's horses in semi-reserves in The Netherlands." *Applied Animal Behavior Science* 33: 303–18.

Keller, Evelyn Fox. 1995. *Refiguring Life: Metaphors of Twentieth-Century Biology*. New York: Columbia University Press.

———. 1998. *Making Sense of Life: Explaining Biological Development with Models, Metaphors, and Machines*. Cambridge: Harvard University Press.

———. 2002. "Competition: Current Usages." In *Keywords in Evolutionary Biology*. Cambridge: Harvard University Press.

King, S.R.B. Behavioural Ecology of Przewalski's Horses reintroduced to Hustai Nuruu National Park, Mongolia. Ph.D. Thesis, University of London, Queen Mary.

King, Sarah R.B. and John Gurnell. 2005. Habitat Use and Spatial Dynamics of Takhi Introduced to Hustai National Park, Mongolia. *Biological Conservation* 124: 277–90.

Kochan, Jeff. 2013. "Subjectivity and Emotion in Scientific Research." *Studies in History and Philosophy of Science* 444: 354–62.

Kohn, Eduardo. 2013. *How Forests Think: Toward an Anthropology Beyond the Human*. Berkeley: University of California Press.

Kolter, L., and W. Zimmerman. 1988. "Social behavior of the Przewalski's horses (*Equus p. przewalskii*) in the Cologne Zoo and its consequences for management and housing." Social contacts between Przewalski horses (*Equus p . Przewalskii*) in Kolner Zoo, in German. *Applied Animal Behavior Science* 21: 117–45.

Kropotkin, Peter. 1972[1902]. *Mutual Aid: A Factor in Evolution*. New York: New York University Press.

Kuhn, Thomas. 1996 [1962]. *The Structure of Scientific Revolutions*. Chicago: University of Chicago Press.

Lai, Whalen. 1981. "Some Notes on the Perceptions of Pratitya-Samutpada in China from Kumarajiva to Fa-Yao." *Journal of Chinese Philosophy* 8: 427–35.

Laski, Harold J. 1919. "The Pluralistic State." *The Philosophical Review* 28(6): 562–75.

Leibniz, Gottfried Wilhelm. 1965. *Monadology and Other Philosophical Essays*. Indianopolis: Bobbs-Merrill Company.

Lefevre, Wolfgang. 2005. "Science as Labor." *Perspectives on Science* 13(2): 194–225.

Leimgruber, Peter, Catherine A. Christen, and Alison Laborderie. 2005. "The Impact of Landsat Satellite Monitoring on Conservation Biology." *Environmental Monitoring and Assessment* 106: 81–101.

Lin, Min. 2001. *Certainty as a Social Metaphor: The Social and Historical Production of Certainty in China and the West*. Westport: Greenwood Press.

Longino, Helen. 1990. *Science as Social Knowledge: Values and Objectivity in Scientific Inquiry*. Princeton: Princeton University Press.

Lysenko, Trofim. 1948. *The Science of Biology Today*. New York: International Publishers.

Macheray, Pierre. 2011. *Hegel or Spinoza?* Susan M. Ruddick, (trans.). Minneapolis: University of Minnesota Press.

Mao Zedong, Deng Xiaoping, and Jiang Zemin. 2008. *Lun Kexue Fazhan*. Theorizing Scientific Development, in Chinese. Beijing: Zhongyang Chubanshe.

Marcus, George E., and Michael M.J. Fischer. 1986. *Anthropology as Cultural Critique: An Experimental Moment in the Human Sciences*. Chicago: University of Chicago Press.

Matherne, Samantha. 2015. "Images and Kant's Theory of Perception." *Ergo: An Open Access Journal of Philosophy* 2(29).

McClusky, Laura. 2001. *Here, Our Culture is Hard: Stories of Domestic Violence from a Mayan Community in Belize*. Austin: University of Texas Press.

Merleau-Ponty, Maurice. 2002[1945]. *The Phenomenology of Perception*. Donald A. Landes. New York: Routledge.

Montaigne, Michel de. 1991. *The Complete Essays*. M.A. Screech (ed.). New York: Penguin.

Mouffe, Chantal. 2013. *Hegemony, Radical Democracy, and the Political*. James Martin (ed). London: Routledge.

Nadasdy, Paul. 2003. *Hunters and Bureaucrats: Power, Knowledge, and Aboriginal-State Relations in Southwest Yukon*. Vancouver: University of British Columbia Press.

Newbold, Tim, Lawrence N. Hudson, Andrew P. Arnell, Sara Contu, Adriana De Palma, Simon Ferrier, Samantha L. L. Hill, Andrew J. Hoskins, Igor Lysenko, Helen R. P. Phillips, Victoria J. Burton, Charlotte W. T. Chng, Susan Emerson, Di Gao, Gwilym Pask-Hale, Jon Hutton, Martin Jung, Katia Sanchez-Ortiz, Benno I. Simmons, Sarah Whitmee, Hanbin Zhang, Jörn P. W. Scharlemann, and Andy Purvis. 2016. "Has Land Use Pushed Terrestrial Biodiversity Beyond the Planetary Boundary? A Global Assessment." *Science* July 15: 288–91.

Nowotny, Helga, Peter Scott, and Michael Gibbons. 2001. *Re-thinking Science: Knowledge and the Public in an Age of Uncertainty*. Cambridge: Polity.

Pascal, Blaise. 1966[1670]. *Pensees*. A. Krailsheimer, translator. New York: Penguin.

Rawls, John. 1971. *A Theory of Justice*. Cambridge: Harvard University Press.

Roberts, Robert C. 2013. *Emotions in the Moral Life*. Cambridge: Cambridge University Press.

Romeijn, Paul, and Edmond Nezry, Pierre Trahan, and Iwan Supit. 2003. "Optical DEM Generation: Satellites Help Preserve Przewalski's Horse." Privateers NV, unpublished report.

Roth, Franziska. 2002. Entwicklung der räumlichen und sozialen Organisation von Przewalskipferden (E. ferus przewalskii) unter naturnahen Bedingungen im Pentezuggebiet (Hortobágy Nationalpark, Ungarn). Development of spatial and social organization of Przewalski horses (E. ferus przewalskii) under natural conditions in the Pentezuggebiet (Hortobágy National Park, Hungary), in German. Ph.D. Dissertation, University of Cologne.

Ryder, Oliver. 1994. "Genetic Studies of Przewalski's Horse and their Impact on Conservation." In *Przewalski's Horse: The History and Biology of an Endangered Species*. Albany: State Univeristy of New York Press.

Sandor, I., and W. Zimmermann. 2000. "Przewalski's Horses in the National Park Hortobágy." *Gazella* 27: 53–64.

Scheffler, Samuel. 1988. *Consequentialism and its Critics*. Oxford: Oxford University Press.

Scott, James C. 1977. *The Moral Economy of the Peasant: Rebellion and Subsistence in Southeast Asia*. New Haven: Yale University Press.

Scott, Michael W. 2007. *The Severed Snake: Matrilineages, Making Place, and a Melanesian Christianity in Southeast Solomon Islands*. Durham: Carolina Academic Press.

Shiel, D., and A. Lawrence. 2004. "Tropical Biologists, Local People, and Conservation: New Opportunities for Collaboration." *Trends in Ecology and Evolution* 19: 634–38.

Silverstein, Michael. 1976. "Shifters, Linguistic Categories, and Cultural Description." In Basso, K. and Selby, H. (eds). *Meaning in Anthropology*, 11–55. Albuquerque: University of New Mexico Press.

Singh, Navindar, and E.J. Milner-Gulland. 2011. "Monitoring Ungulates in Central Asia: Current Constraints and Future Potential." *Oryx* 45(1): 38–49.

Skiff, E.M. 1982. The effect of enclosure design on social interactions and daily activity patterns of the captive Asiatic wild horse (*Equus Przewalskii*). M.S. Thesis, University of Minnesota.

Spalton, J. A., M. W. Lawrence, and S.A. Brend. 1999. "Arabian Oryx Reintroduction in Oman: Successes and Setbacks." *Oryx* 33(2): 168–75.

Strathern, Marilyn. 1991. *Partial Connections*. Lanham, MD: Rowman & Littlefield.

———. 1992[1989]. *After Nature: English Kinship in the Late Twentieth Century*. Cambridge: Cambridge University Press.

Thompson, E.P. 1971. "The Moral Economy of the English Crowd in the Eighteenth Century." *Past & Present* 50 (Feb): 76–136.

Tsing, Anna Lowenhaupt. 2015. *The Mushroom at the End of the World: On the Possibility of Life in Capitalist Ruins*. Princeton: Princeton University Press.

Turner, J. 2010. "Ontological Pluralism." *Journal of Philosophy* 107: 5–34.

Van Dierendonck, M.C., N. Bandi, D. Batdorj, S. Dugerlham, and B. Munkhtsog. 1996. "Behavioral observations of the reintroduced takhi or Przewalski's horses (*Equus ferus przewalski*.) in Mongolia." *Applied Animal Behavior Science* 50: 95–114.

Viveiros de Castro, Eduardo Batalha. 2014. *Cannibal Metaphysics: For a Post-Structural Anthropology*. Minneapolis: Univocal.

Walzer, C., and P. Kaczensky. 2008. "Radio Waves, Explosions and Satellites—Choosing the Appropriate Telemetry Device for Wildlife Studies." Egzotikus es Attatkerti Emtosdttatok Betegsegei, March 14–16, Budapest, Hungary, unpublished paper.

Wenger, Etienne. 2000. "Communities of Practice and Learning Systems." *Organization* 7(2): 225–46.

Wit, Piet, and Inge Bouman. 2006. *The Tale of the Przewalski's Horse: Coming Home to Mongolia*. Utrecht: KNNV Publishers.

Wittgenstein, Ludwig. 1968[1953]. *Philosophical Investigations*. Oxford: Blackwell.

Zhang, Feng. 2009. Pushi Yema Xingwei Shengtai Jiqi Xiangguan Yinzide Yanjiu. The Behavioral Ecology and Related Factors in the Study of Przewalski's Horses, in Chinese. Beijing Forestry University. MA Thesis.

Index

About the Author

Michael L. Zukosky is an associate professor of anthropology. He has general teaching interests in cultural and linguistic anthropology as well as in Asian Studies. His research interests are in history of anthropology, philosophical anthropology, the intersection of environmental anthropology and the anthropology of science and technology, and international development and conservation policy. The latter interests involve applied work in social assessment in both Asia and the United States.